Selectivity in Organic Synthesis

Selectivity in Organic Synthesis

Robert S. Ward
University of Wales Swansea

JOHN WILEY & SONS
Chichester • New York • Weinheim • Brisbane • Singapore • Toronto

Other Wiley Editorial Offices

John Wiley & Sons, Inc., 605 Third Avenue,
New York, NY 10158-0012, USA

WILEY-VCH Verlag GmbH, Pappelallee 3,
D-69469 Weinheim, Germany

Jacaranda Wiley Ltd, 33 Park Road, Milton,
Queensland 4064, Australia

John Wiley & Sons (Asia) Pte Ltd, 2 Clementi Loop #02-01,
Jin Xing Distripark, Singapore 129809

John Wiley & Sons (Canada) Ltd, 22 Worcester Road,
Rexdale, Ontario M9W IL1, Canada

Library of Congress Cataloging-in-Publication Data

Ward, Robert S.
      Selectivity in organic synthesis / Robert Ward.
         p.   cm.
      Includes bibliographical references and index.
      ISBN 0-471-98778-6 (alk. paper).– ISBN 0-471-98779-4 (pbk. :
      alk. paper)   1. Organic compounds—Synthesis.   2. Stereochemistry.   I. Title.
   QD262.W27   1999
   547′.2—dc21   98-44894
                                                                     CIP

British Library Cataloging in Publication Data

A catalogue record for this book is available from the British Library

ISBN 0 471 98778 6 (HB)
ISBN 0 471 98779 4 (PB)

Typeset in 10/12pt Times by Laser Words, Madras, India

Contents

Preface

This book is based upon a 10 lecture course given to final year undergraduate and first year postgraduate students over a number of years. It is partly as a result of its format as a lecture course that aspects of selectivity involved in the chemistry of carbonyl compounds are dealt with in two chapters and similarly the reactions of alkenes are also dealt with in two separate chapters. Nevertheless, despite the fact that each of the chapters had its origins in a one hour lecture slot, it will be immediately obvious that there is more than enough material in most of the chapters for at least two lectures. Thus, when lecturing on a topic of such active, current interest it is very easy to find new and more elegant examples to illustrate each topic. This obviously has the advantage that different themes can be emphasized and different examples given each time the course is presented. The course is never the same from one year to the next. In any event the material accumulated falls neatly into nine chapters. A problem session is an essential part of any lecture course dealing with advanced organic chemistry and with this in mind a number of problems with answers and explanations are included at the end of each chapter.

The aim in producing this book has been to provide an introductory text aimed at final year undergraduates and postgraduates. It is intended to be concise, readable and interesting. No attempt has been made to give a comprehensive coverage of the literature on any topic but suggested references for further reading are presented at the end of each chapter.

Although it would be easy to see this as just another book on asymmetric synthesis an attempt has been made to treat selectivity in a more general context. For this reason the synthesis of chiral non-racemic compounds starting from the so-called 'chiral pool' is not an aspect which receives explicit attention, although the use of chiral auxiliaries is referred to in several places and several of the syntheses quoted do indeed start from naturally occuring chiral compounds. Similarly, although several examples of enzyme catalysed reactions are included, the importance of biocatalysts is not given separate consideration. The selectivity involved in biochemical reactions is obviously an enormous topic in its own right. Rather, the emphasis has been placed on aspects of selectivity, including chemoselectivity and regioselectivity, which are inherent in organic reactions, and the strategies that can be adopted to emulate or even surpass the high selectivity found in most biological reactions.

Finally I should like to express my thanks to Dr Martyn Pritchard of the Parke-Davis Neuroscience Research Centre, Cambridge, and Dr Christine Willis of the University of Bristol, for reading and commenting on the draft manuscript and making a number of helpful suggestions.

Robert S. Ward

Abbreviations

A,B,C, D	unspecified (spectator) groups not involved in the reaction being illustrated or unspecified compounds used for illustration purposes
acac	acetylacetone (2,4-pentanedione)
Ar*	aryl group containing an element of chirality
9-BBN	9-borabicyclo[3.3.1]nonane
BINAL-H	lithium (1,1'-binaphthyl-2,2'-dioxy)ethoxy aluminium hydride
BINAP	2,2'-bis(diphenylphosphino)-1,1'-binaphthyl
BMS	borane-dimethyl sulfide
Bn	benzyl
Boc	*tert*-butoxycarbonyl
Bz	benzoyl
cat*	chiral catalyst
CB	catecholborane
COD	1,5-cyclooctadiene
CSA	camphorsulfonic acid
Δ	heat
DCCI	*N,N'*-dicyclohexylcarbodiimide
d.e.	diastereomeric excess
DEAD	diethyl azodicarboxylate
DET	diethyl tartrate
DHQ	dihydroquinine
DHQD	dihydroquinidine
DIBAL	diisobutylaluminium hydride
DIOP	2,3-*O*-isopropylidene-2,3-dihydroxy-1,4-bis(diphenylphosphino)butane = 4,5-bis(diphenylphosphinomethyl)-2,2-dimethyl-dioxolane
DIPAMP	1,4-bis[*o*-methoxyphenyl(phenyl)phosphino]ethane

DIPT	diisopropyl tartrate
DMAP	4-(*N*,*N*-dimethylamino)pyridine
DMSO	dimethylsulfoxide
DPPB	1,4-bis(diphenylphosphino)butane
E$^+$	electrophile
e.e.	enantiomeric excess
equiv.	molar equivalent
HOMO	highest occupied molecular orbital
HMPA	hexamethylphosphoric triamide
Ipc	isopinocampheyl
KHMDS	potassium hexamethyldisilazide
L	ligand
L*	chiral ligand
LDA	lithium diisopropylamide
LHMDS	lithium hexamethyldisilazide
LUMO	lowest unoccupied molecular orbital
M	metal (including B, Si, etc.)
MAD	methyl aluminium bis(2,6-di-*tert*-butyl-4-methylphenoxide)
MCPBA	*meta*-chloroperbenzoic acid
MOM	methoxymethyl
Ms	methanesulfonyl
NaHMDS	sodium hexamethyldisilazide
NBS	*N*-bromosuccinimide
Nu$^-$	nucleophile
P	product
PCC	pyridinium chlorochromate
PDC	pyridinium dichromate
PHAL	phthalazine
PPTS	pyridinium *p*-toluenesulfonate
pyr	pyridine
R$_S$, R$_M$, R$_L$	small, medium and large groups
R*	alkyl or aryl group containing stereogenic centre or element of chirality
RAMP	(*R*)-1-amino-2-methoxymethylpyrrolidine
Ra-Ni	Raney nickel
Red-Al	sodium bis(2-methoxyethoxy)aluminium hydride
S	substrate (starting material)
SAMP	(*S*)-1-amino-2-methoxymethylpyrrolidine
TBAF	tetrabutylammonium fluoride
TBS	*tert*-butyldimethylsilyl
TBDPS	*tert*-butyldiphenylsilyl
TBHP	*tert*-butyl hydroperoxide

Tf	trifluoromethanesulfonyl
TFA	trifluoroacetic acid
THF	tetrahydrofuran
THP	tetrahydropyranyl
TIPS	triisopropylsilyl
TMEDA	N, N, N', N'-tetramethylethylenediamine
TMS	trimethylsilyl
Tol	*p*-tolyl (4-methylphenyl)
Tr	trityl (triphenylmethyl)
Ts	*p*-toluenesulfonyl
X	group containing heteroatom (O, N, Cl, etc.) or group which can be readily modified or replaced
X*	chiral auxiliary group

1

What do We Mean by Selectivity?

The whole basis of organic chemistry, and especially organic synthesis, depends upon the selectivity which can be achieved in organic reactions. Selectivity can be defined as the discrimination observed in a reaction involving competitive attack on two or more substrates or at two or more positions, groups or faces in the same substrate. Several different kinds of selectivity can be identified, and as a result different levels of control can be exercised over the outcome of organic reactions. This book describes the strategies which can be adopted to improve selectivity and reactions which have been specially designed to afford high selectivity. The aim is to illustrate the range of processes to which these principles can be applied and the high degree of selectivity which can be achieved. Consider, first of all, two alternative situations in which selectivity arises.

1.1 REACTIONS INVOLVING DISCRIMINATION BETWEEN DIFFERENT SUBSTRATES

This can be described as *substrate selectivity* and involves a reagent which transforms different substrates, e.g. A and B, under the same conditions, at different rates, into the products C and D.

$$A \xrightarrow[k_1]{\text{reagent}} C$$

$$k_1 \neq k_2$$

$$B \xrightarrow[k_2]{\text{reagent}} D$$

The two substrates, A and B, can be structural isomers as in the preferential hydrogenation of 1-hexene in the presence of its tetrasubstituted isomer (eqn. 1.1), and the preferential reaction of *tert*-butanol with concentrated hydrochloric acid in the presence of its primary isomer (eqn. 1.2).

$$(1.1)$$

$$(1.2)$$

The two substrates can also be diastereomers as in the oxidation of *cis*- and *trans*-4-*tert*-butylcyclohexanol (eqn. 1.3), and the elimination reactions of the isomeric 2,3-dibromobutanes (eqn. 1.4).

$$(1.3)$$

$$k_1 > k_2 \qquad (1.4)$$

Discrimination between enantiomers leads to kinetic resolution (eqns 1.5 and 1.6). This topic is discussed in more detail in Chapter 7.

(1.5)

| racemic | 59% 63% e.e.* | 37% 95% e.e.* |

*Enantiomeric excess (e.e.) = major enantiomer (%) − minor enantiomer (%)

(1.6)

1.2 REACTIONS INVOLVING DISCRIMINATION BETWEEN DIFFERENT SITES IN THE SAME SUBSTRATE

This can be described as *product selectivity* and involves a reaction in which more than one product can be formed but in which the products are formed in a ratio which differs from the statistically expected one.

$$A \xrightarrow{\text{reagent}} B + C + D \dots \qquad [B] \neq [C] \neq [D] \dots$$

For example, discrimination between different positions in a molecule can lead to the preferential formation of one (or more) structural isomers (eqn. 1.7). The statistically expected ratio in this particular case would be 2:2:1, if each of the five sites on the benzene ring were equally reactive. This type of selectivity is defined as *regioselectivity* (Section 1.3).

(1.7)

Discrimination between different groups or faces can lead to the preferential formation of one (or more) diastereomers (eqns 1.8 and 1.9). This type of selectivity is defined as *diastereoselectivity* (Section 1.3).

$$\text{(1.8)}$$

1. BH_3/THF
2. H_2O_2 ^-OH

$$\text{(1.9)}$$

1. EtMgBr
2. H_3O^+

3 : 1

In some situations discrimination between different groups or faces leads to the preferential formation of one enantiomer (eqns 1.10 and 1.11). This type of selectivity is defined as *enantioselectivity* (Section 1.3).

$$\text{(1.10)}$$

H_2O
fumarase
(an enzyme)

$$\text{(1.11)}$$

PhCHO + BuLi

Ph Bu 40% e.e.

In most situations the requirement is to discriminate between different sites in the same molecule. Apart from kinetic resolution (see Chapter 7), this will therefore be the main theme in the chapters which follow.

1.3 DEFINITIONS OF SELECTIVITY

The very simplest, the most obvious, and probably the most important type of selectivity is *chemoselectivity*, which can be defined as the discrimination shown by a reagent in its reaction with different functional groups, i.e. its preferential reaction with one functional group in a molecule containing two (or more) functional groups (eqns 1.12 and 1.13).

$$\text{(1.12)}$$

$NaBH_4$

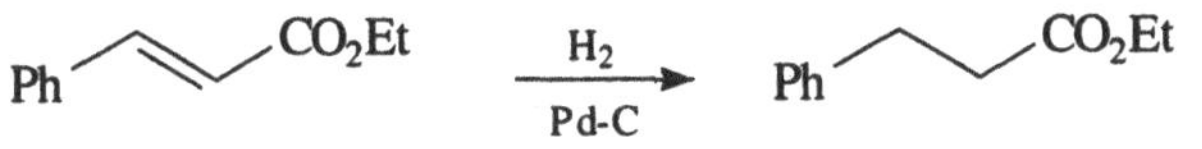

$$Ph\diagup\diagdown\diagup CO_2Et \xrightarrow[\text{Pd-C}]{H_2} Ph\diagup\diagdown\diagup\diagdown CO_2Et \qquad (1.13)$$

We take chemoselectivity for granted. Where would organic chemistry be without it?

Regioselectivity can be defined as the preferential reaction at one (or more) possible sites in a molecule resulting in the preferential formation of one (or more) structural isomers (eqn. 1.7). An example involving conjugate addition, in which the degree of selectivity varies with the nature of the aryl group, is shown in Scheme 1.1.

Stereoselectivity can be defined as the preferential formation of one (or more) products that differ only in their configuration. Stereoselectivity can be further subdivided into enantioselectivity and diastereoselectivity.

Enantioselectivity occurs when the two stereoisomeric products which can be formed are enantiomers (eqns. 1.10, 1.11, 1.14 and 1.15).

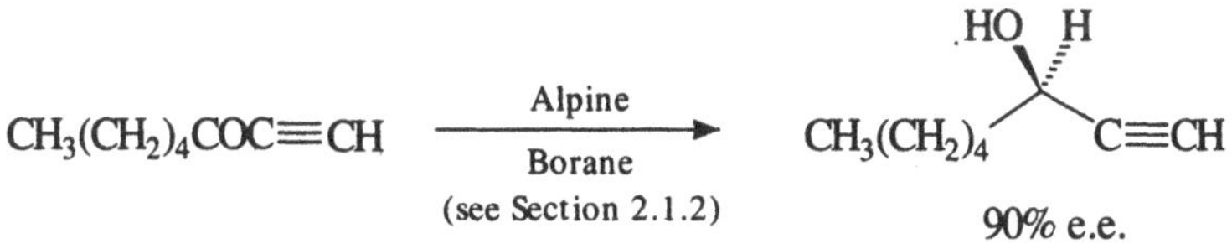

$$CH_3(CH_2)_4COC\equiv CH \xrightarrow[\substack{\text{Borane}\\ \text{(see Section 2.1.2)}}]{\text{Alpine}} \underset{\text{90\% e.e.}}{CH_3(CH_2)_4 \overset{\overset{\displaystyle HO\ H}{\diagup}}{\diagdown} C\equiv CH}$$

$$(1.14)$$

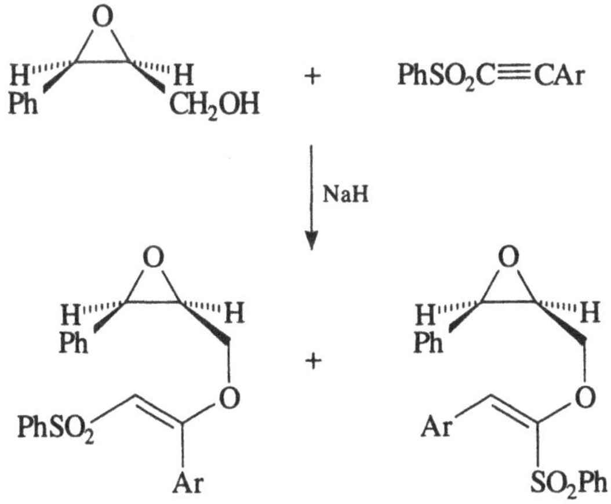

Ar = phenyl	51	:	49
Ar = 3,4–dimethoxyphenyl	86	:	14
Ar = 3,4–methylenedioxyphenyl	88	:	12

Scheme 1.1

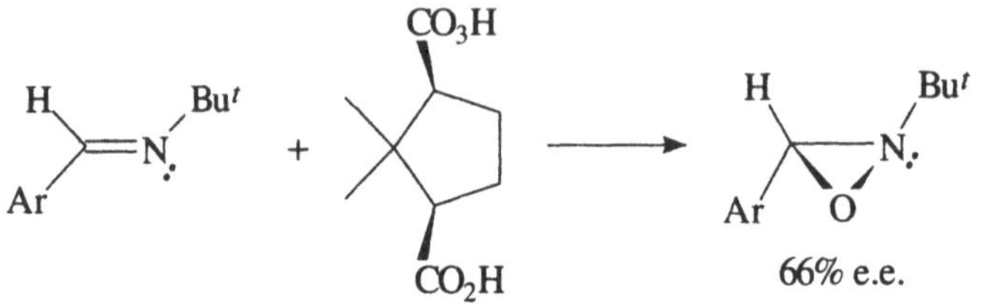

$$(1.15)$$

66% e.e.

Diastereoselectivity occurs when the products which can be formed are diastereomers. Diastereoselectivity can be further subdivided into two types, which for want of a better term are referred to as simple and absolute diastereoselectivity. *Simple* diastereoselectivity can occur in any reaction in which two or more new stereogenic centres are created, even when the reaction involves an achiral substrate and an achiral reagent (1.8, 1.16 and 1.17). The product will be either achiral or racemic.

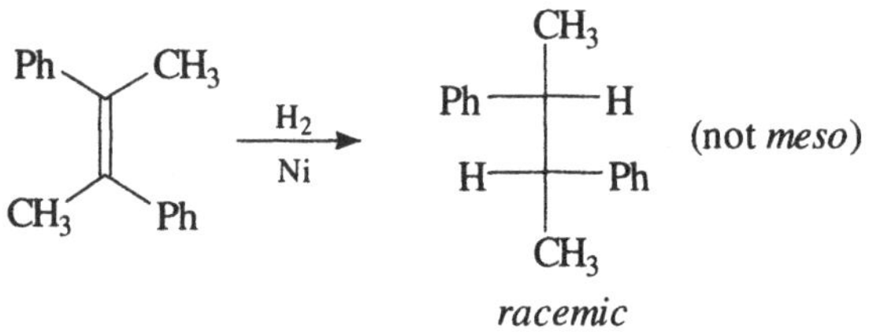

$$(1.16)$$

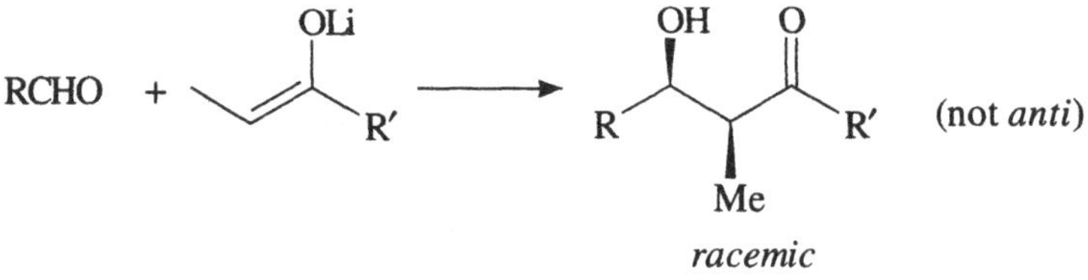

racemic

$$(1.17)$$

racemic

Absolute diastereoselectivity is seen in the reaction of a chiral substrate with an achiral reagent (eqns 1.9, 1.18 and 1.19). If the starting material is non-racemic then the product will also be non-racemic.

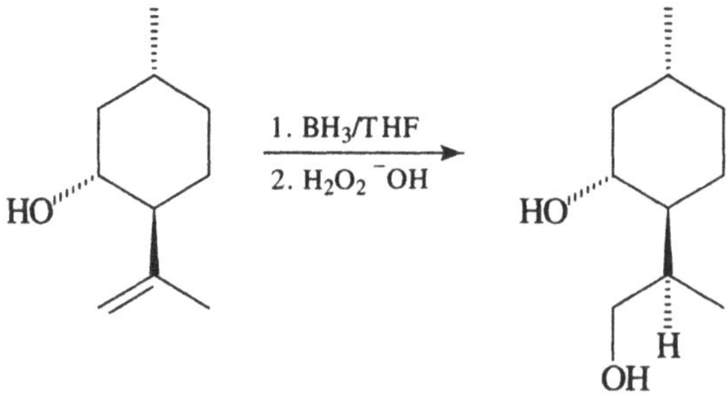

$$(1.18)$$

$$\text{(1.19)}$$

α-pinene

It is of course possible for a reaction to involve both enantioselectivity and diastereoselectivity (eqns 1.20 and 1.21).

$$\text{(1.20)}$$

$$\text{(1.21)}$$

1.4 STEREOTOPIC AND STEREOFACIAL SELECTIVITY

If one of the faces of a molecule is attacked preferentially then the reaction is said to involve facial selectivity. New stereogenic centres can also be created by discrimination between two apparently identical atoms or groups. The terms *homotopic*, *enantiotopic* and *diastereotopic* are used to describe the stereochemical relationship between groups or faces in a molecule. Thus, groups (or atoms) are said to be *homotopic* if they can be interchanged by rotation about an axis of symmetry. Similarly, the two faces of a double bond are homotopic if the plane dividing them contains an axis of symmetry. Transformations of homotopic groups, or addition to homotopic faces, always give rise to the same product (eqns 1.22 and 1.23).

$$\text{(1.22)}$$

$$\text{(1.23)}$$

Two groups are *enantiotopic* if they are related to one another by a plane or centre of symmetry. Similarly, two faces are enantiotopic if the plane dividing them is a plane of symmetry which does not contain a coplanar axis of symmetry. Transformations of enantiotopic groups, or addition to enantiotopic faces, usually give rise to enantiomers (eqns 1.24 and 1.25). Reactions which involve preferential attack on one of the two enantiotopic groups or faces are said to exhibit *enantiotopic or enantiofacial* selectivity (e.g. eqns 1.10 and 1.11). Such reactions can be brought about by the use of chiral reagents or chiral catalysts.

$$(1.24)$$

$$(1.25)$$

Two groups which cannot be interchanged by any symmetry operation are described as *diastereotopic*. Similarly, two faces are diastereotopic if the plane dividing them is not a plane of symmetry and does not contain an axis of symmetry. Transformations of diastereotopic groups, or addition to diastereotopic faces, give rise to diastereomers (eqns 1.26 and 1.27). Reactions which involve preferential attack by a reagent on one of the two diastereotopic groups or faces are said to exhibit *diastereotopic or diastereofacial* selectivity (e.g. eqns 1.18 and 1.19).

$$(1.26)$$

$$(1.27)$$

The stereochemical relationship between the various terms defined in this section is further illustrated in Figure 1.1.

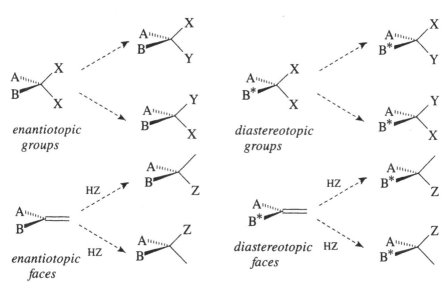

Figure 1.1

PROBLEMS

1. Identify the types of selectivity operating in each of the following reactions:

(a)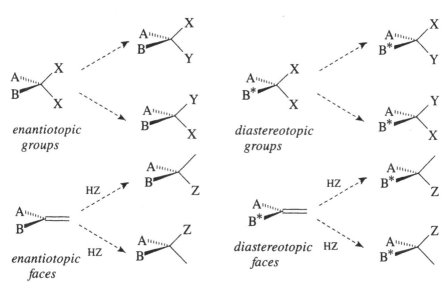

(b)

racemic endo (3 : 1) racemic exo

(c)

68%   32%

2. Identify the stereotopic and stereofacial selectivity involved in each of the following reactions:

(a)

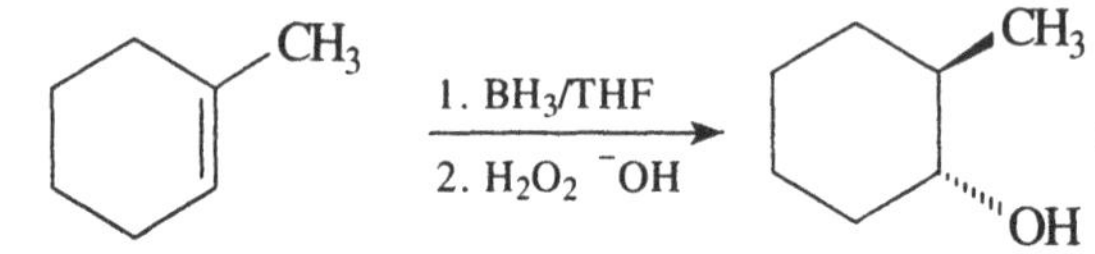

$$\text{1. BH}_3/\text{THF}$$
$$\text{2. H}_2\text{O}_2 \ \ \overline{\ }\text{OH}$$

major diastereomer
(racemic)

(b)

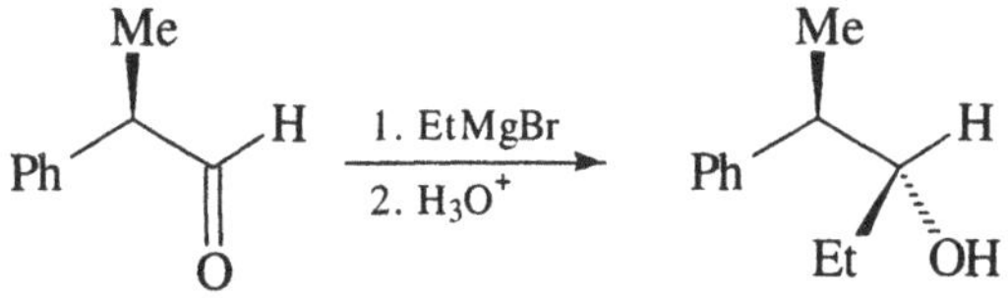

Me, Ph, H, O

1. EtMgBr
2. H$_3$O$^+$

Me, Ph, H, Et, OH

major diastereomer

(c) 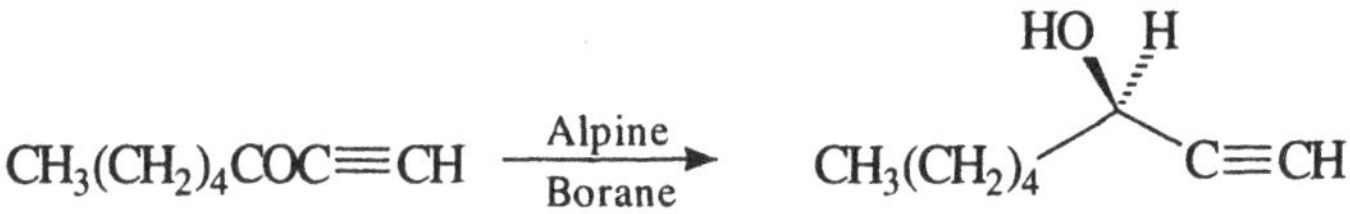 $\text{CH}_3(\text{CH}_2)_4\text{COC}{\equiv}\text{CH}$

$$\xrightarrow[\text{Borane}]{\text{Alpine}}$$

HO H
$\text{CH}_3(\text{CH}_2)_4$ $\text{C}{\equiv}\text{CH}$

90% e.e.

(d)

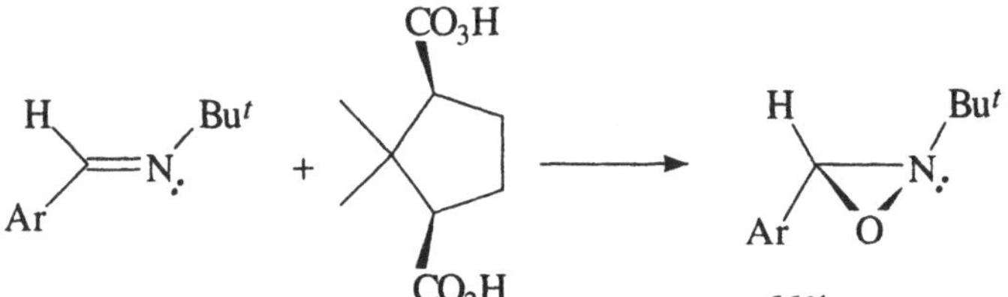

H, Bu^t
=N.
Ar

+

CO$_3$H

CO$_2$H

⟶

H, Bu^t
-N.
Ar O

66% e.e.

(e)

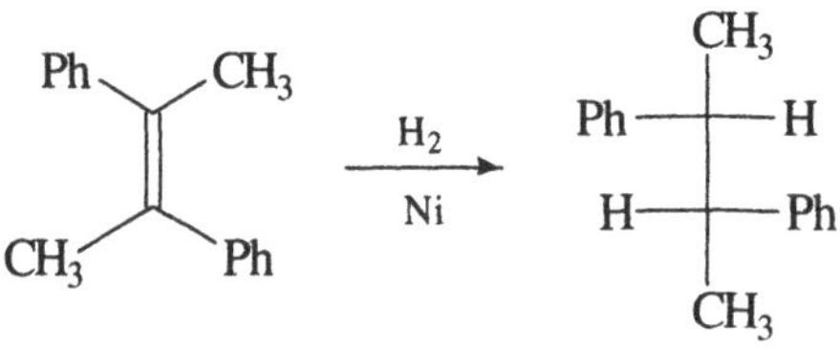

Ph, CH$_3$
CH$_3$, Ph

$$\xrightarrow[\text{Ni}]{\text{H}_2}$$

CH$_3$
Ph—H
H—Ph
CH$_3$

major diastereomer
(racemic)

(f)

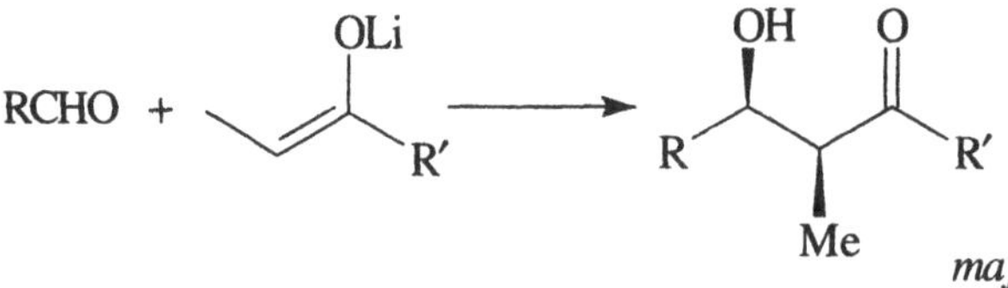

RCHO +

OLi
R′

⟶

OH   O
R   R′
Me

major diastereomer
(racemic)

(g)

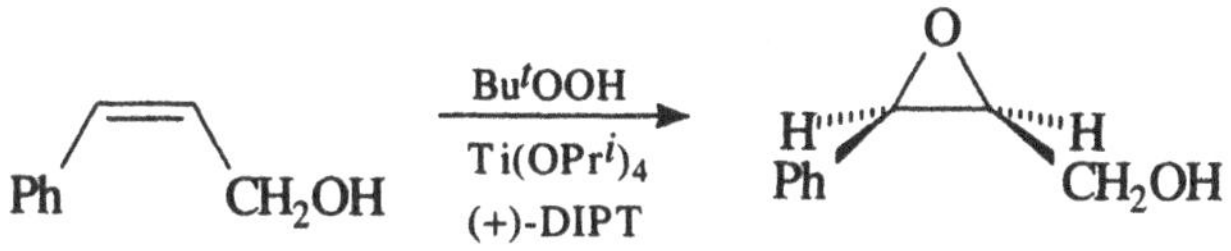

major stereoisomer

(h)

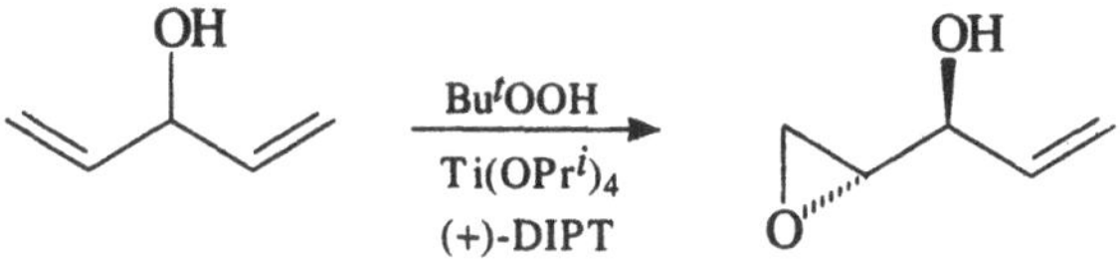

major stereoisomer

FURTHER READING

V. Gold, *Pure Appl. Chem.*, 1983, *55*, 1281.
M. Nogradi, *Stereoselective Synthesis*, 2nd edition, VCH, Weinheim, 1995, chap. 1.
M. Maier, in *Organic Synthesis Highlights II* (ed. H. Waldemann), VCH, Weinheim, 1995.
R. S. Ward, *Chemistry in Britain*, 1991, 803.
A. Pelter, R. S. Ward and G. M. Little, *J. Chem. Soc., Perkin Trans. 1*, 1990, 2775.

ANSWERS

1. (a) enantioselectivity, enantiofacial selectivity.
   (b) diastereoselectivity
   (c) diastereoselectivity, diastereofacial selectivity.
2. (a) none
   (b) diastereofacial selectivity
   (c) enantiofacial selectivity
   (d) enantiofacial selectivity
   (e) none
   (f) none
   (g) enantiofacial selectivity
   (h) enantiotopic and diastereofacial selectivity (see Section 6.4 and Chapter 7).

2

Stereoselective Reactions of Carbonyl Compounds I

Reactions of carbonyl compounds are dealt with in this chapter and in Chapter 4. Broadly speaking, nucleophilic addition (excluding the aldol reaction) is dealt with in this chapter and enolate reactions (including the aldol reaction) are dealt with in Chapter 4.

2.1 NUCLEOPHILIC ADDITION TO CARBONYL COMPOUNDS

There are three ways of controlling the stereoselectivity of addition to carbonyl compounds:

- (i) use of a chiral substrate;
- (ii) use of a chiral reagent;
- (iii) use of a chiral catalyst.

2.1.1 Use of a chiral substrate

If the substrate is chiral then two diastereomeric transition states occur with the result that two diastereomeric products are formed in unequal amounts (eqns 2.1 and 2.2).

$$\begin{array}{ll} R = Me & 40\% \text{ d.e.}^* \\ R = Et & 50\% \text{ d.e.} \\ R = Ph & 60\% \text{ d.e.} \end{array}$$

(2.1)

$$\begin{array}{c} \text{CH}_3 \\ \text{H} \end{array}\!\!\!\!\!\!\overset{\displaystyle O}{\underset{\displaystyle \text{Ph}}{\bigg\backslash\!\!\!\!\bigg/}}\!\!\!\!\text{Me} \quad \xrightarrow[\text{2. H}_3\text{O}^+]{\text{1. LiAlH}_4} \quad \begin{array}{c} \text{CH}_3 \\ \text{H} \end{array}\!\!\!\!\!\!\overset{\displaystyle \text{OH}}{\underset{\displaystyle \text{Ph}}{\bigg\backslash\!\!\!\!\bigg/}}\!\!\!\!\overset{}{\underset{\displaystyle \text{R}}{\text{Me}}}$$

$$\quad (2.2)$$

R = Me	50% d.e.*
R = Et	50% d.e.
R = Pr^i	66% d.e.
R = Bu^t	96% d.e.

*Diastereomeric excess (d.e.) = major diastereomer (%) – minor diastereomer (%)

Several rules have been developed to account for the stereoselectivity of such reactions, the most well known being attributed to Cram (Figure 2.1). Cram's rule predicts that the nucleophile will attack from the side of the smallest group when the molecule adopts a conformation in which the largest group is *syn* coplanar to the carbonyl group. Cram's model does not accurately represent either the ground state or the transition state of the reaction but it does provide a useful method for predicting the structure of the major product. A better mechanistic rationalization is provided by the Felkin model, which is also supported by *ab initio* calculations. Felkin's rule considers two conformations and predicts that the major one, or at least the one leading to the major product, will be the one shown on the left-hand side of the equilibrium in Figure 2.1.

The above rules work well when only steric factors are involved. In such situations the degree of selectivity which can be achieved is usually modest. The above rules do not apply when chelation occurs. In this situation greater selectivity is usually achieved (eqns 2.3–2.5), and the outcome of such reactions

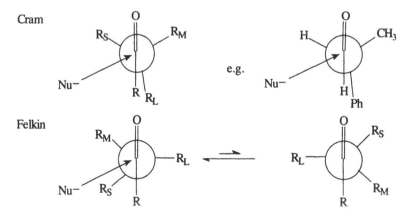

Figure 2.1

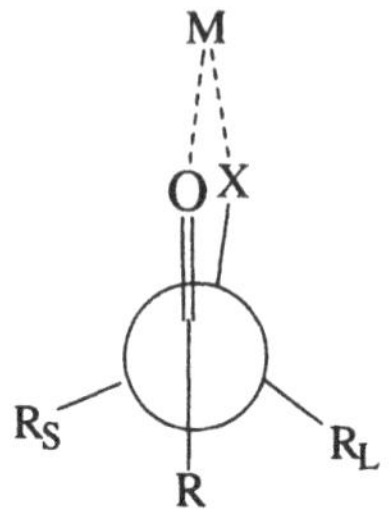

Cram's chelate
(cyclic) model

Figure 2.2

can be accounted for by assuming that the group containing a heteroatom on the adjacent carbon atom is held *syn* coplanar to the carbonyl group. This is sometimes referred to as Cram's chelate (cyclic) model (Figure 2.2).

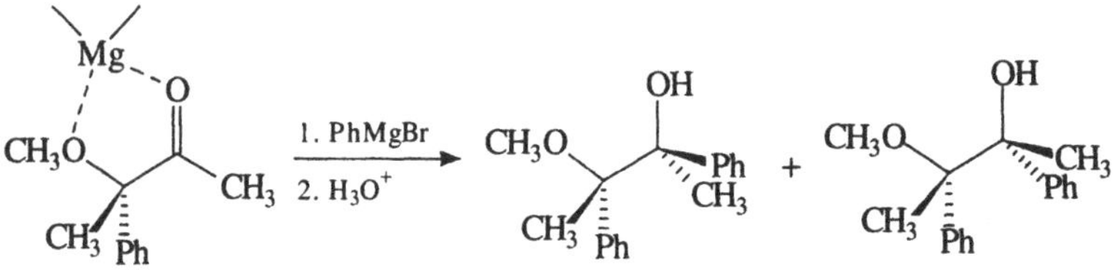

$$86 \quad : \quad 14$$

$$(2.3)$$

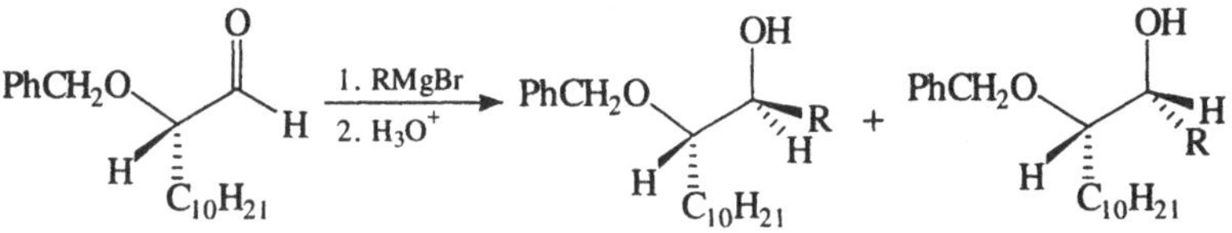

$$R = Ph(CH_2)_3 \qquad 94 \quad : \quad 6$$
$$R = CH_2 = CH(CH_2)_3 \qquad 95 \quad : \quad 5$$

$$(2.4)$$

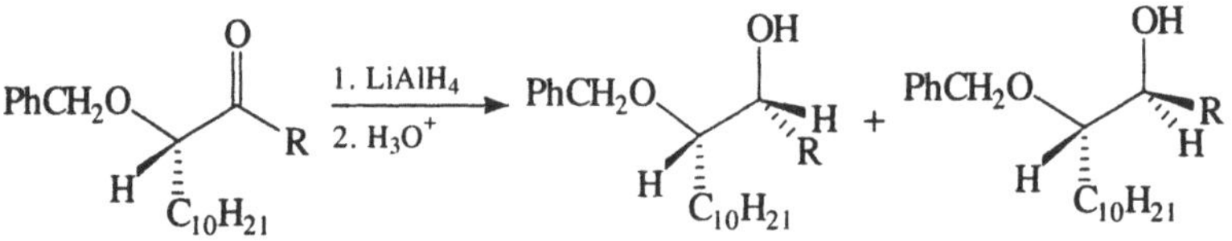

$$R = Ph(CH_2)_3 \qquad 76 \quad : \quad 24$$
$$R = CH_2 = CH(CH_2)_3 \qquad 80 \quad : \quad 20$$

$$(2.5)$$

The examples shown above all depend upon using a stereogenic centre already present in the substrate in order to construct a new stereogenic centre. An alternative approach is to incorporate a chiral auxiliary group into the substrate which can be later removed. This approach is illustrated by the elegant work of Whitesell *et al.*, who used 8-phenylmenthol as the chiral auxiliary (eqns 2.6 and 2.7). This approach has been used to synthesize both enantiomers of the pheromone frontalin (Schemes 2.1 and 2.2).

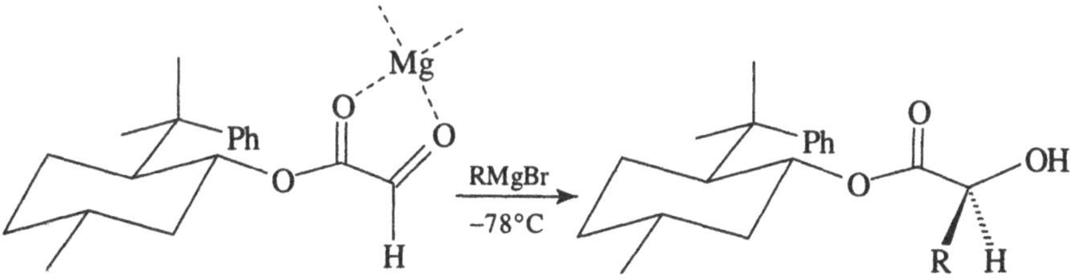

$$R = Me \quad 98\% \text{ d.e.}$$

$$(2.6)$$

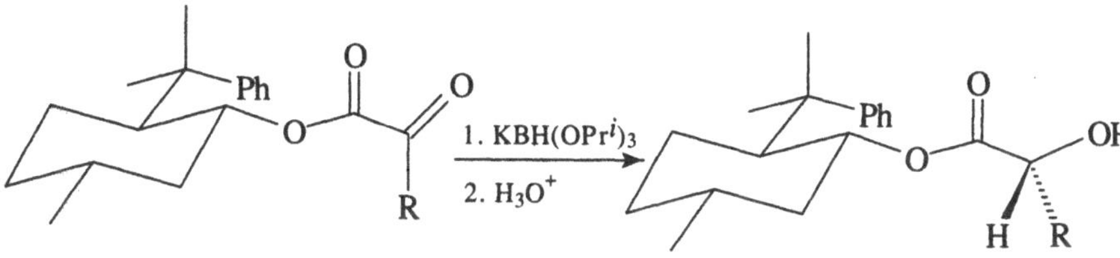

$$R = Me \quad 90\% \text{ d.e.}$$

$$(2.7)$$

2.1.2 Use of a chiral reagent

Several chiral reducing agents have been developed. For example, (*R*)- and (*S*)-BINAL-H are derivatives of lithium aluminium hydride which contain a homochiral binaphthol unit (Scheme 2.3). A model based on a cyclic chair-like transition state has been proposed to account for the observed stereoselectivity. The favoured arrangement places the larger substituent equatorial and the smaller one axial. The use of (*S*)-BINAL-H in the first step of a synthesis of a chiral butyrolactone is illustrated in Scheme 2.3.

A number of borane derived reducing reagents have also been developed. Alpine-borane and chlorodiisopinocampheylborane are both derived from α-pinene. Once again a transition state model can be used to rationalize the observed stereoselectivity of the reduction. Alpine-borane is particularly effective for the asymmetric reduction of aldehydes, β-dicarbonyl compounds, α,β-unsaturated ketones, and acetylenic ketones. Its low reactivity towards other ketones can be overcome by increasing the Lewis acidity of the boron.

(single diastereoisomer)

1. LiAlH$_4$
2. H$_3$O$^+$

O$_3$
$-78°C$

(−)-frontalin
(100% e.e.)

Scheme 2.1

This has been achieved by the development of chlorodiisopinocampheylborane (Scheme 2.4).

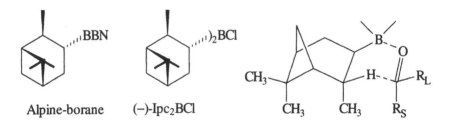

Alpine-borane   (−)-Ipc$_2$BCl

BBN = 9-borabicyclo[3.3.1]nonyl   transition state model

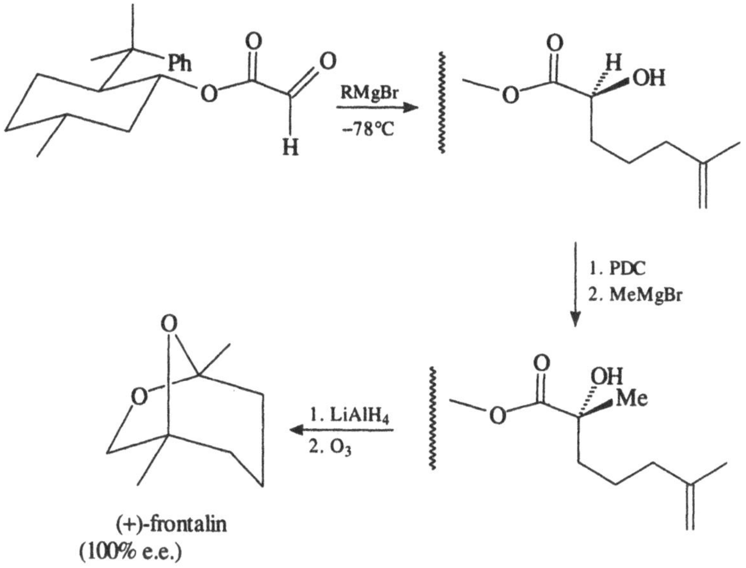

Scheme 2.2

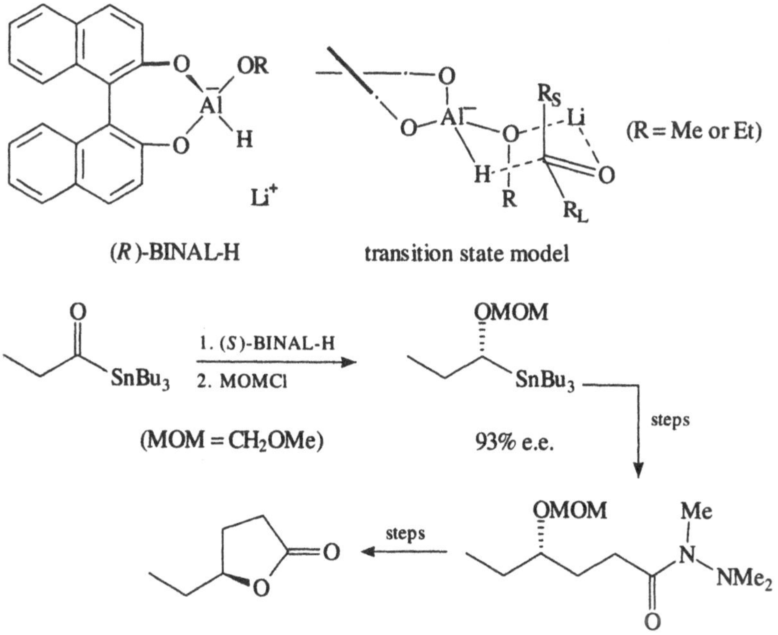

(*R*)-BINAL-H   transition state model

Scheme 2.3

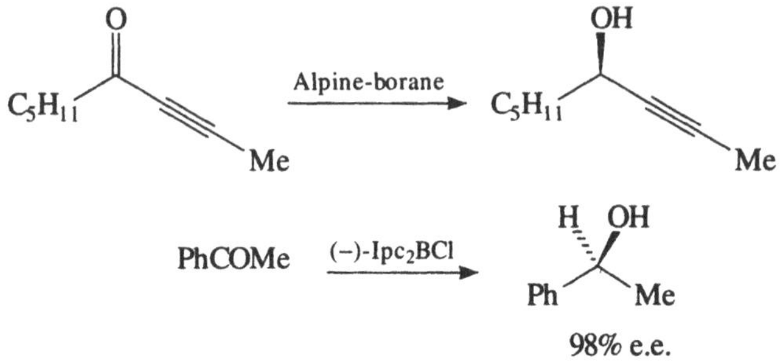

Scheme 2.4

2.1.3 Use of a chiral catalyst

A major advance has been the development of the oxazaborolidine catalysts by Corey *et al.* (Scheme 2.5). They are prepared from proline and are unique in that they contain a Lewis acidic site and a Lewis basic site in close proximity which activate both the carbonyl group and the borane simultaneously. The transition state model has been corroborated by X-ray analysis of the borane adduct and by quantum mechanical calculations. The catalytic cycle involving the oxazaborolidine is illustrated in Scheme 2.6.

Other nucleophilic addition reactions which have been extensively studied include addition of butyllithium and diethylzinc to benzaldehyde (Schemes 2.7 and 2.8). Several chiral ligands have been shown to enhance the enantioselectivity of these reactions.

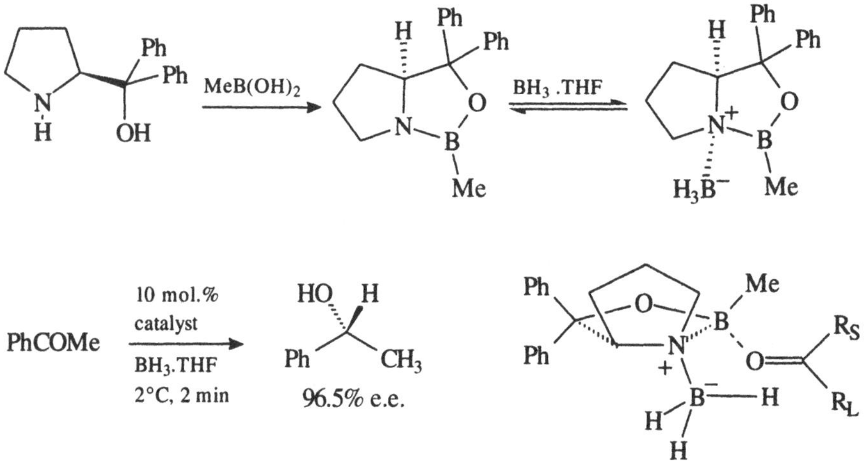

Scheme 2.5

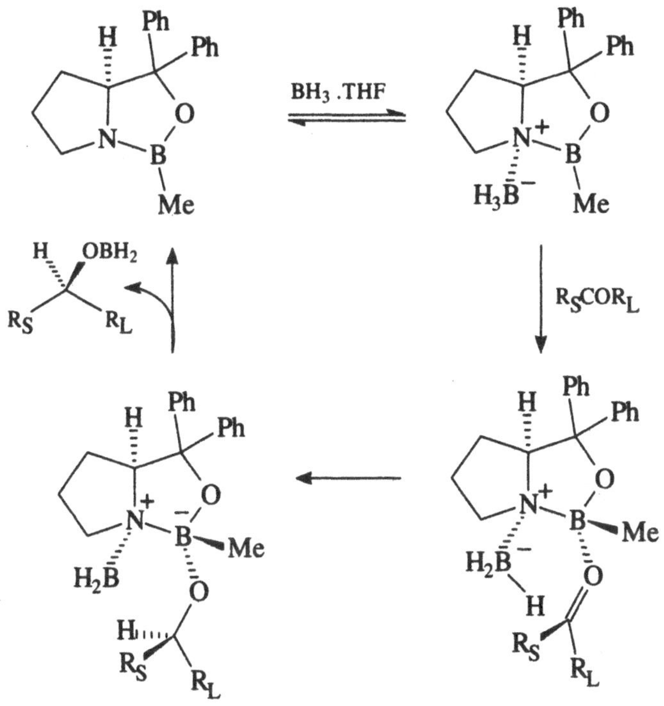

Scheme 2.6

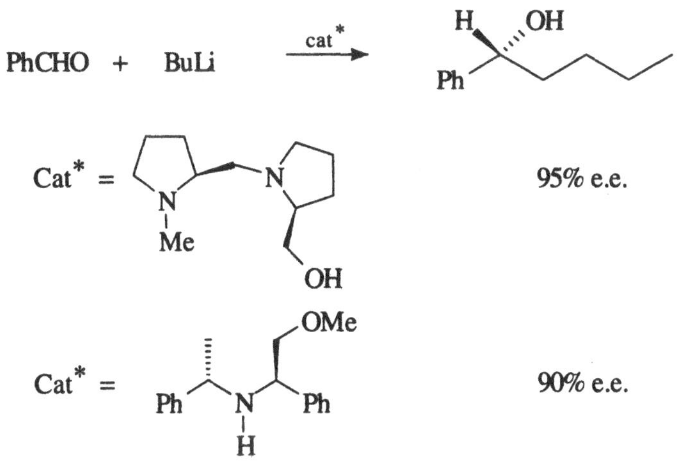

Scheme 2.7

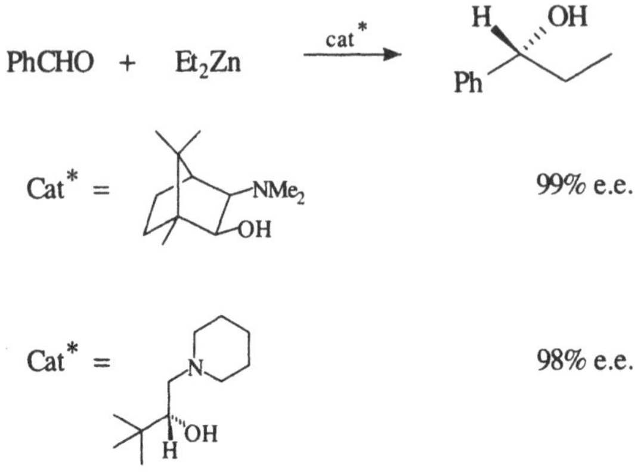

$$\text{PhCHO} \;+\; \text{Et}_2\text{Zn} \xrightarrow{\text{cat}^*} \underset{\text{Ph}}{\overset{\text{H} \quad \text{OH}}{\diagdown}}$$

Cat* = [structure with NMe₂ and OH]   99% e.e.

Cat* = [piperidine structure with "OH and H]   98% e.e.

Scheme 2.8

2.2 ASYMMETRIC CONJUGATE ADDITION

Nucleophilic addition to $\alpha\beta$-unsaturated carbonyl compounds can lead to the generation of a new chiral centre at the β position. Furthermore, when the intermediate enolate is trapped by an electrophile there is the potential to create new chiral centres at both the α and β positions (eqn. 2.8). The relative and/or absolute configuration at these positions can in principle be controlled using the same techniques as those applicable to 1,2-addition to the carbonyl group.

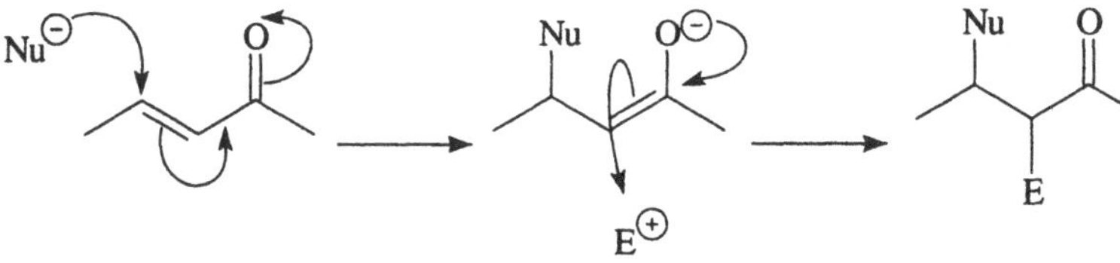

$$\text{(2.8)}$$

For example, tandem conjugate addition to cyclic substrates leads to a *trans* arrangement of the two incoming groups (eqn. 2.9). Furthermore when an adjacent chiral centre is present in the molecule the incoming nucleophile enters *trans* to the group already present (eqns 2.10 and 2.11).

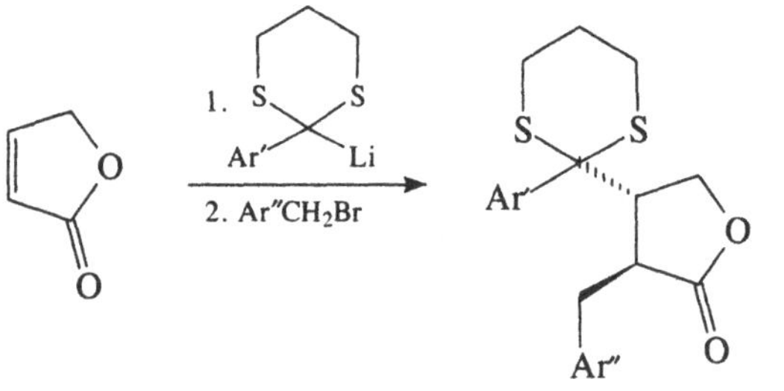

$$\text{(2.9)}$$

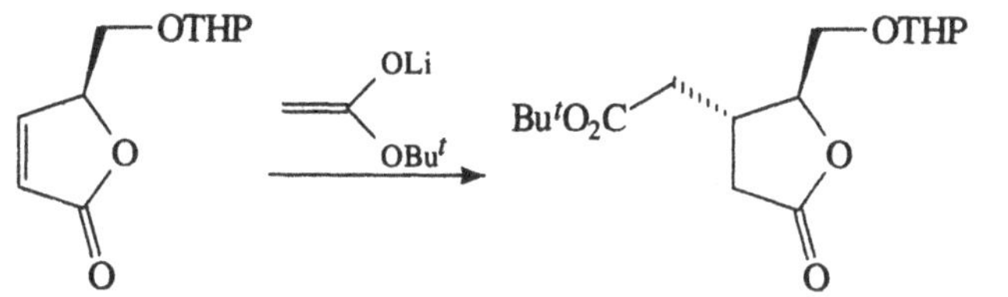

$$(2.10)$$

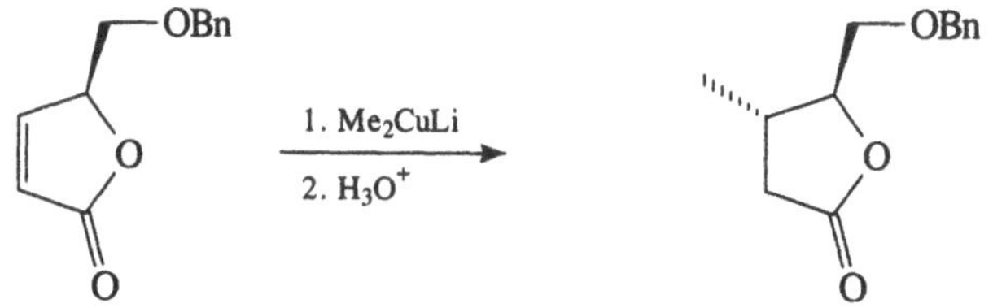

$$(2.11)$$

Two examples in which the configuration at the β position is determined by a chiral sulfoxide are shown in Scheme 2.9 and eqn. 2.12.

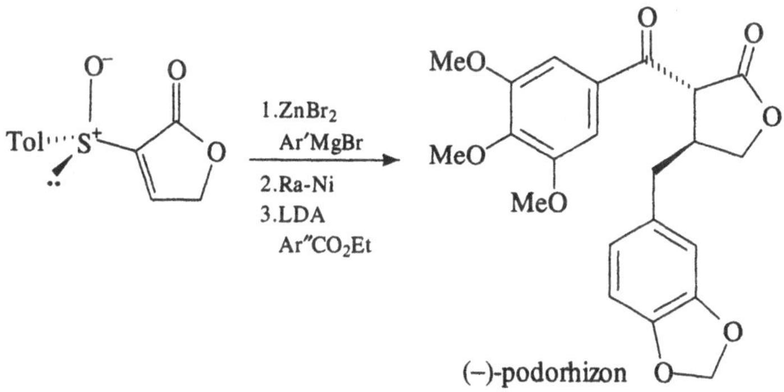

$$(2.12)$$

(−)-podorhizon

Furthermore, by using nucleophiles and electrophiles other than alkyl groups, a variety of α- and β-substituted carbonyl compounds, including α- and β-amino acids, can be prepared (eqns 2.13 and 2.14 and Scheme 2.10).

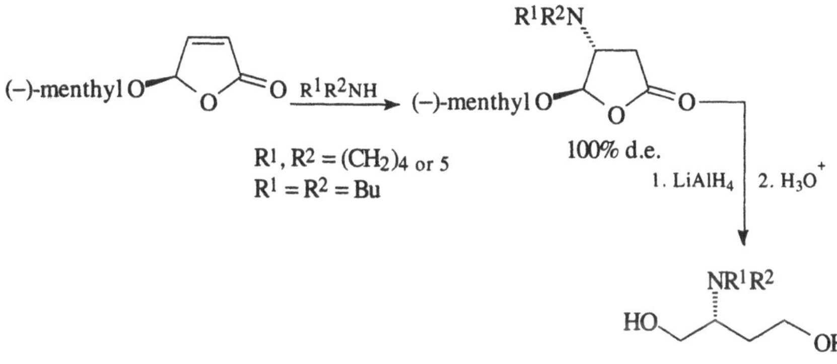

$R^1, R^2 = (CH_2)_4$ or 5
$R^1 = R^2 = Bu$

100% d.e.

1. LiAlH$_4$ 2. H$_3$O$^+$

$$(2.13)$$

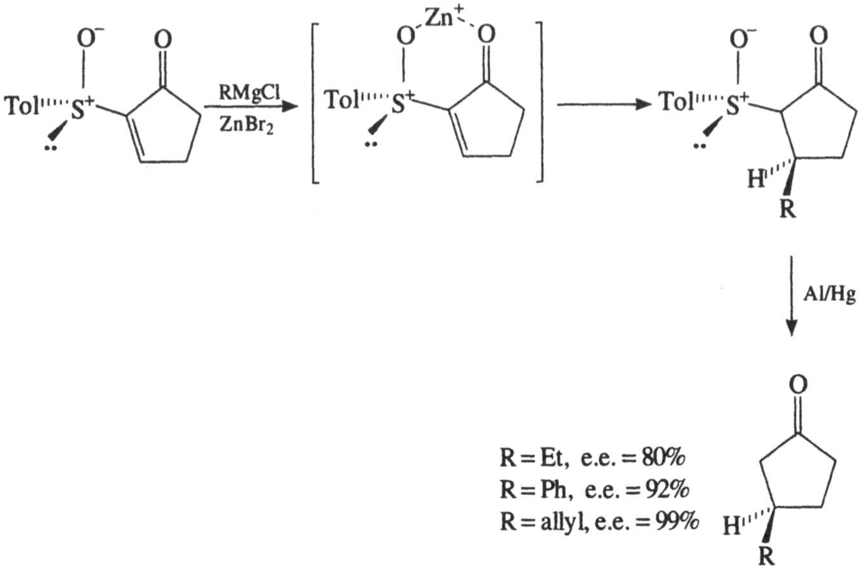

$R = Et$, e.e. $= 80\%$
$R = Ph$, e.e. $= 92\%$
$R = $ allyl, e.e. $= 99\%$

Scheme 2.9

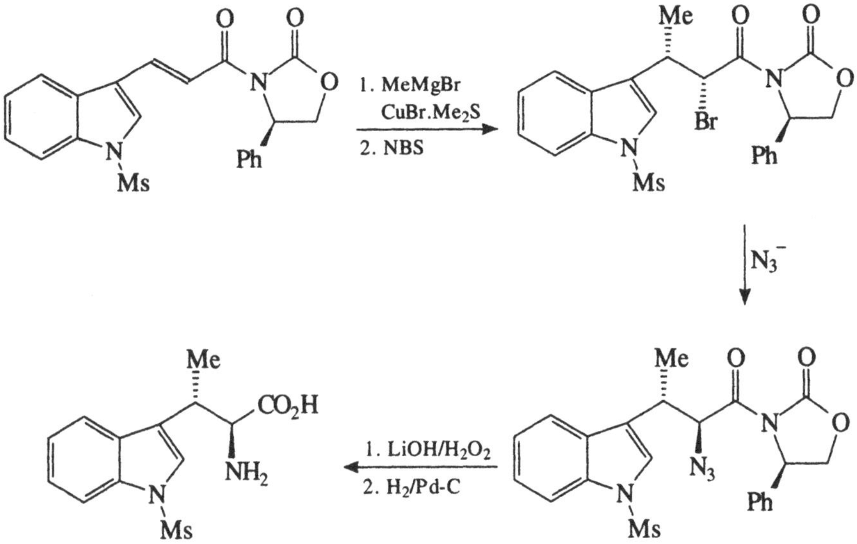

Scheme 2.10

(2.14)

2.3 ADDITION OF ALLYLBORON DERIVATIVES

Chiral boron reagents have a wide variety of uses in organic synthesis. Thus, they provide versatile reagents for the stereoselective reduction of ketones (Section 2.1.2) and as enolate derivatives they can be utilized in alkylation and aldol reactions (see Chapter 4). A useful reaction of allylborane reagents is their reaction with aldehydes (eqn. 2.15). The boron atom in the leaving group is able to coordinate with the non-bonding pair of the carbonyl group to form a six-membered cyclic transition state. The product is an alkoxyborane which is readily hydrolysed to give the corresponding alcohol.

(2.15)

The relative configuration (*syn* or *anti*) of the product is determined by the configuration (*E* or *Z*) of the carbon–carbon double bond in the allyl group (eqns 2.16 and 2.17). The terms *syn* and *anti* are used to denote whether the OH and the Me group are both on the same side (i.e. both above or both below the plane of the paper) when the carbon chain is written in an extended (zigzag) form, or on opposite sides (i.e. one above and one below), respectively.

(2.16)

anti (racemic)

(2.17)

syn (racemic)

By having chiral groups attached to boron it is possible to selectively obtain a single enantiomer of either the *syn* or the *anti* isomer (eqns 2.18–2.21). Even when a stereogenic centre is present in the aldehyde the configuration of the groups on boron determine the absolute configuration of the product.

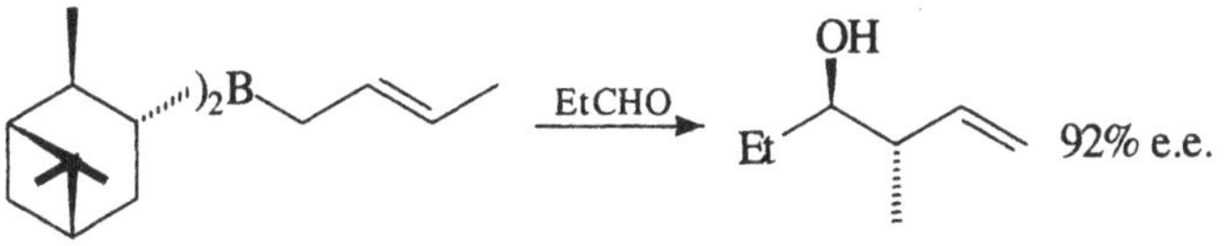

(2.18)

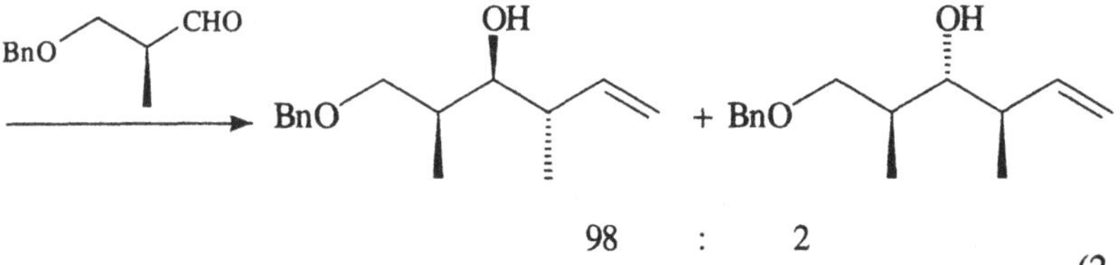

98 : 2

(2.19)

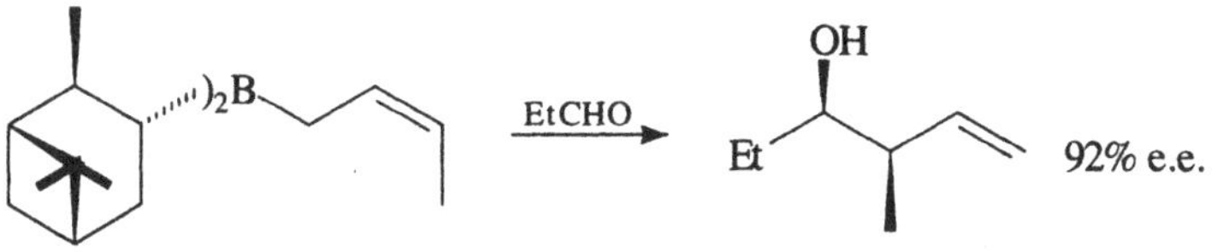

(2.20)

92 : 8

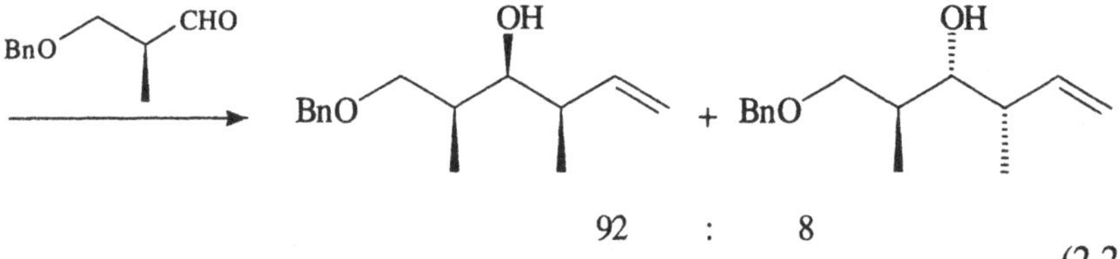

(2.21)

Similar results have also been obtained using other chiral allyl boron derivatives (eqns 2.22–2.25).

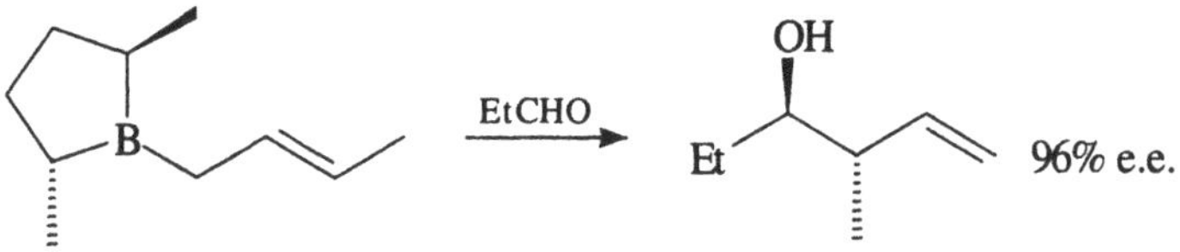

(2.22)

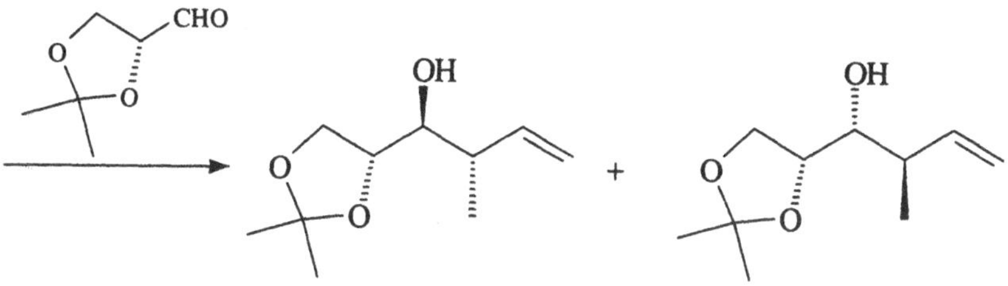

(2.23)

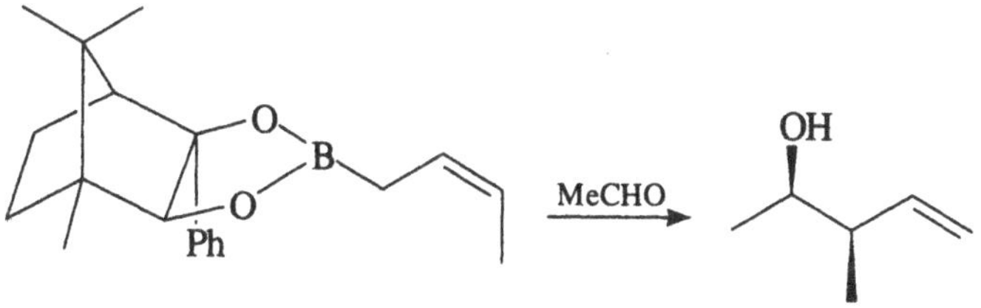

70% e.e.

(2.24)

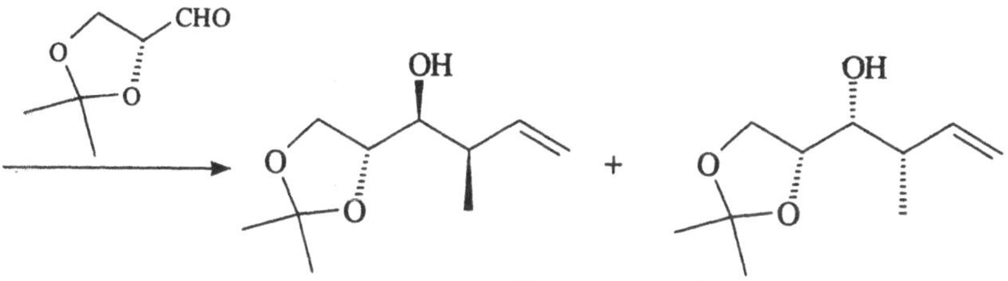

(2.25)

PROBLEMS

1. Using either Cram's rule or Felkin's rule rationalize the following transformation:

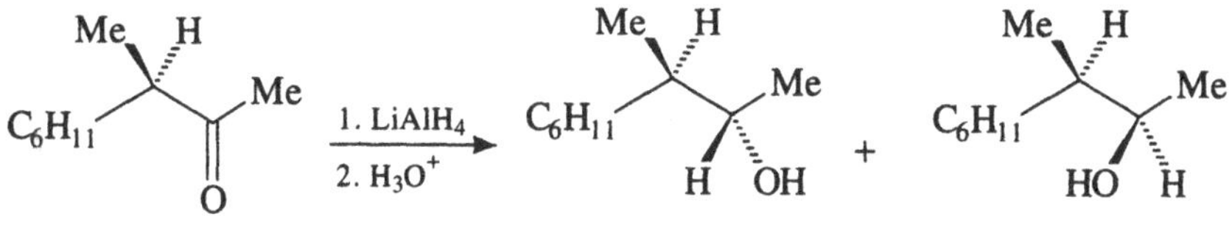

major product   minor product

2. The alcohols **3** and **4** can be prepared in a stereoselective manner by addition of Grignard reagents to the ketones **1** and **2**. Account for the stereoselectivity of the reactions.

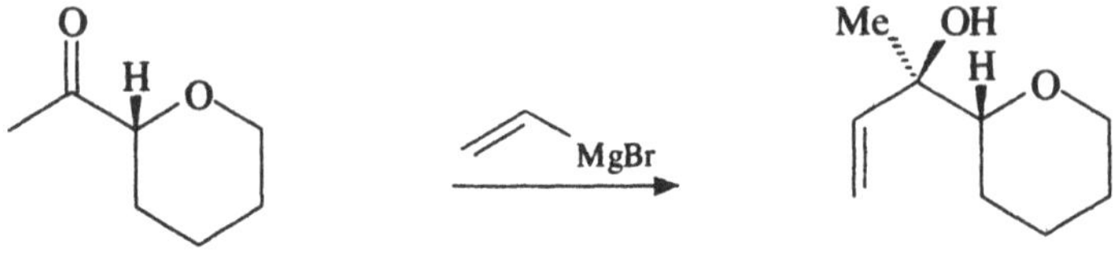

1 **3**

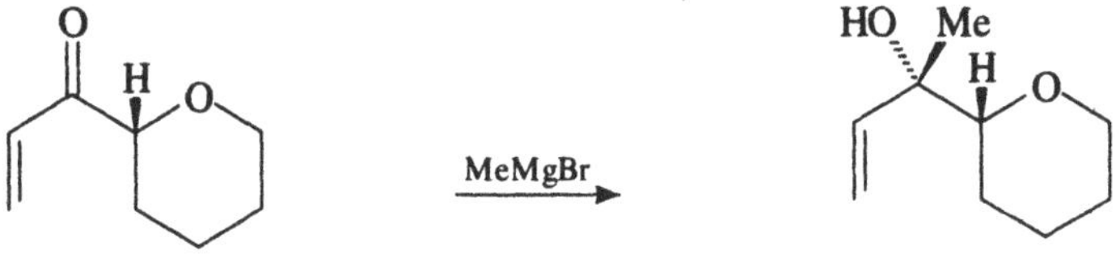

2 **4**

3. By studying eqns 2.22–2.25 predict the major stereoisomer formed in each of the following reactions:

(a)

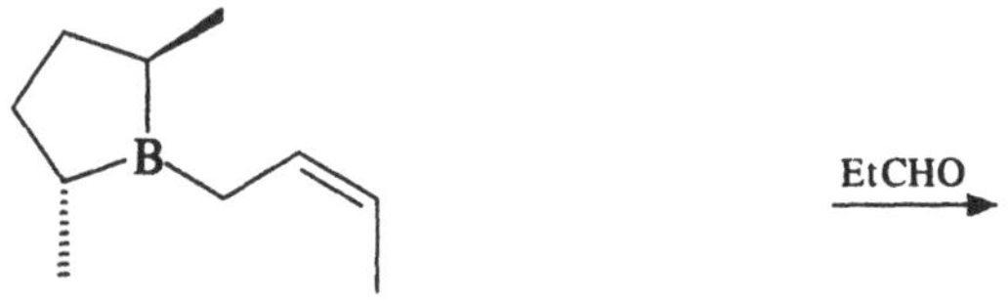

(b)

(c)

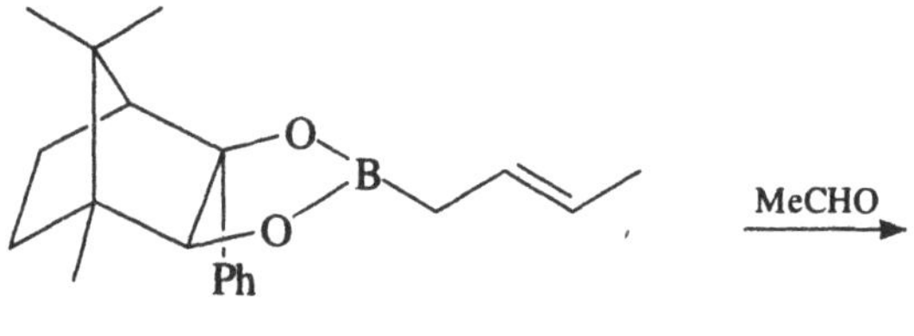

(d)

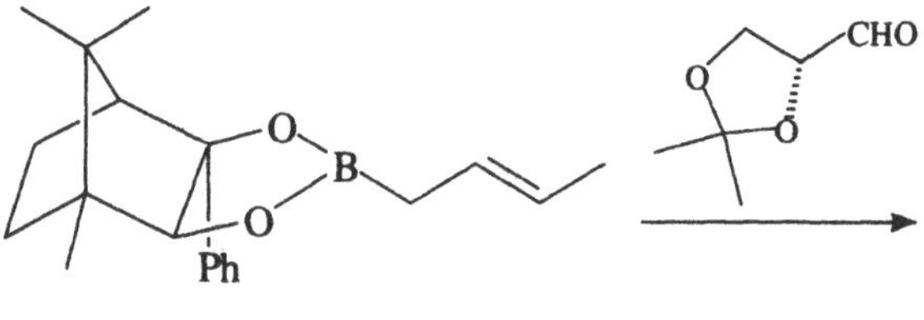

4. The boronic esters **5** and **6** react with the chiral aldehyde **7** to give the products indicated. Predict the major products formed by reacting **5** and **6** with the enantiomer of **7**.

93 : 5

14 : 85

FURTHER READING

A. Koskinen, *Asymmetric Synthesis of Natural Products*, Wiley, Chichester, 1993, chap. 3.

Cram's rule

J. M. Fleischer, A. J. Gushurst and W. L. Jorgensen, *J. Org. Chem.*, 1995, *60*, 490.

Chelation controlled addition

M. T. Reetz, *Angew. Chem. Int. Ed.*, 1984, *23*, 556.
J. K. Whitesell, *Accts Chem. Res.*, 1985, *18*, 280.
J. K. Whitesell and C. M. Buchanan, *J. Org. Chem.*, 1986, *51*, 5443.

Chlorodiisopinocampheylborane

J. Chandrasekharan, P. V. Ramachandran and H. C. Brown, *J. Org. Chem.*, 1985, *50*, 5446.

Asymmetric conjugate addition

B. E. Rossiter and N. M. Swingle, *Chem. Rev.*, 1992, *92*, 771.
J. Leonard, *Contemporary Organic Synthesis*, 1995, *1*, 387.

Addition by allylboron reagents

H. C. Brown and K. S. Bhat, *J. Am. Chem. Soc.*, 1986, *108*, 5919.

H. C. Brown, K. S. Bhat and R. S. Randad, *J. Org. Chem.*, 1989, *54*, 1570.

J. Garcia, B.-M. Kim and S. Masamune, *J. Org. Chem.*, 1987, *52*, 4831.

R. W. Hoffmann, H. J. Zeiss, W. Ladner and S. Tabche, *Chem. Ber.*, 1982, *115*, 2357.

W. R. Roush, K. Ando, D. B. Powers, A. D. Palkowitz and R. L. Halterman, *J. Am. Chem. Soc.*, 1990, *112*, 6339.

W. R. Roush, A. D. Palkowitz and K. Ando, *J. Am. Chem. Soc.*, 1990, *112*, 6339 and 6348.

ANSWERS

1.

Cram   Felkin

2.

3.

(a)

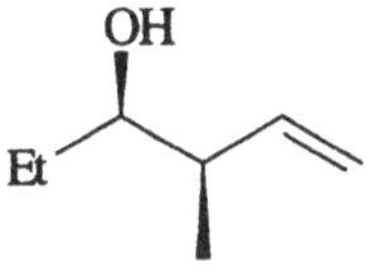

(b)

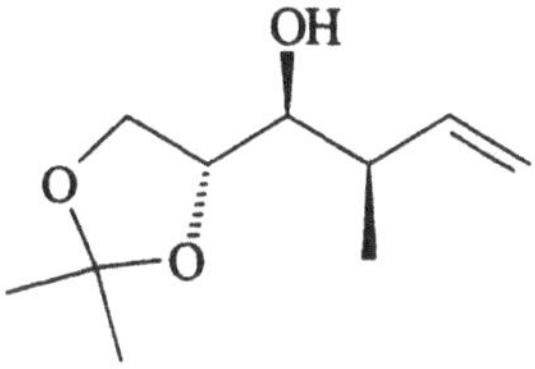

(c)

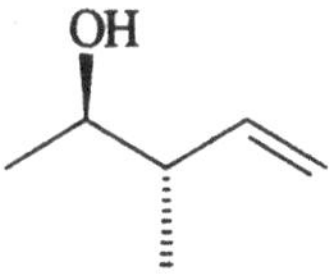

(d)

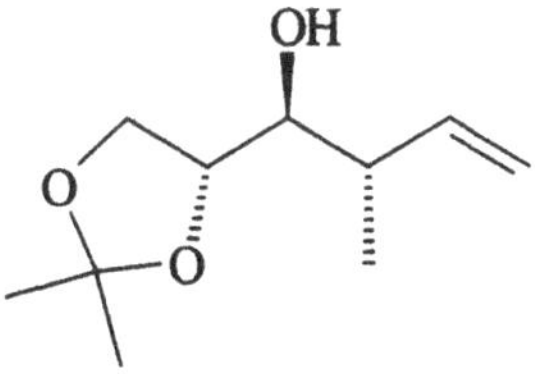

4.

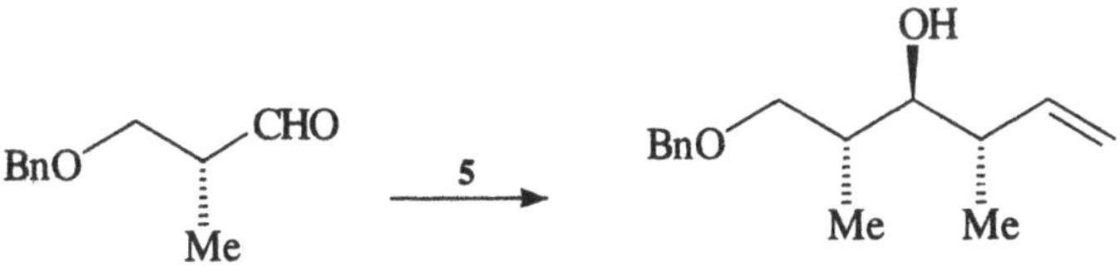

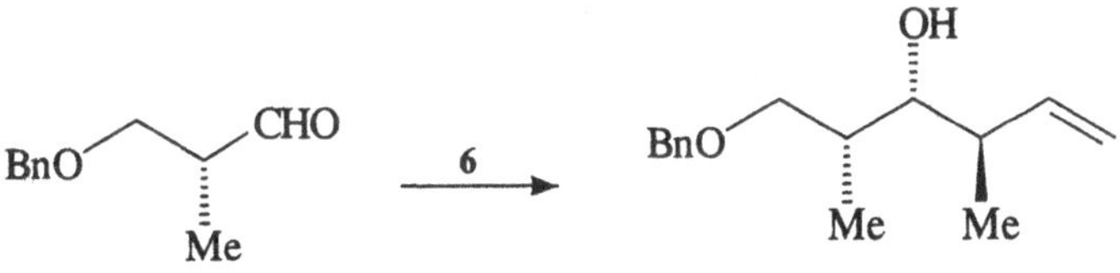

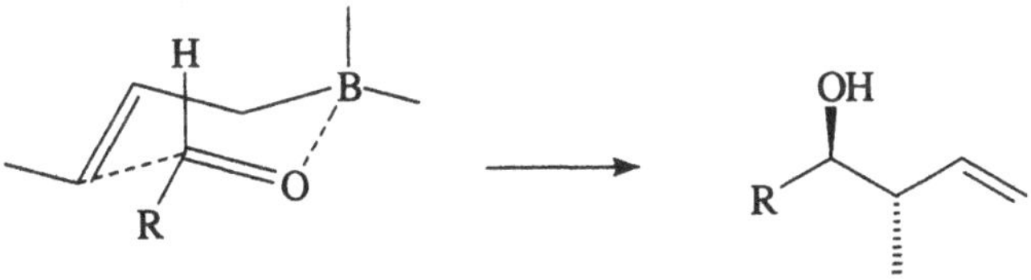

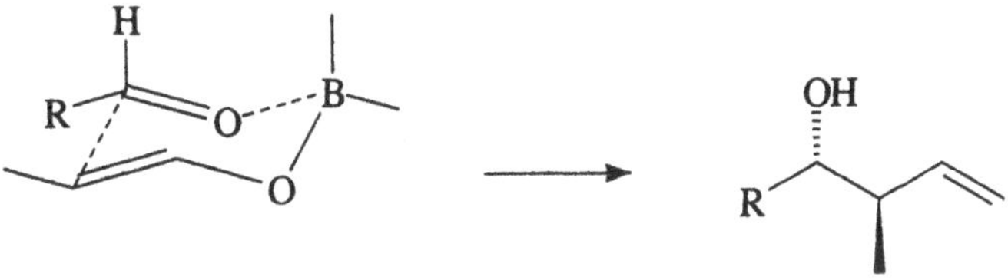

3

Stereoselective Reactions of Alkenes I

Treatment of the stereoselective reactions of alkenes is divided between this chapter and Chapter 5. Broadly speaking, the pericyclic reactions of alkenes are dealt with in this chapter and epoxidation, dihydroxylation, cyclopropanation, etc., are dealt with in Chapter 5.

3.1 THE DIELS ALDER REACTION

The immense value of the Diels Alder reaction is due to its high regio- and stereoselectivity. Besides forming two new carbon–carbon bonds, the reaction generates up to four new stereogenic centres. The fact that the reaction proceeds through a highly ordered cyclic transition state exercises precise control over the configuration of the new stereogenic centres.

Consider first the *regioselectivity* of the reaction. The reaction of an unsymmetrical diene with an unsymmetrical dienophile can, in principle, lead to two isomeric adducts but in practice one usually predominates. Thus, in the addition of acrylic acid derivatives to 1-substituted butadienes the *"ortho"* (1,2-) adduct is favoured (eqn. 3.1), while with 2-substituted butadienes the *"para"* (1,4-) isomer predominates (eqn. 3.2). These orientation effects are governed largely by the orbital coefficients at the termini of the two conjugated systems. The atoms with the larger orbital coefficients bond preferentially in the transition state. In most cases this leads to the 1,2- (*"ortho"*) adduct with 1-substituted butadienes and to the 1,4- (*"para"*) adduct with 2-substituted butadienes. In the presence of Lewis acids, the polarization of the dienophile is increased by coordination to the Lewis acid and this leads to an increase in regioselectivity. Under these conditions very high yields of a single isomer can often be obtained.

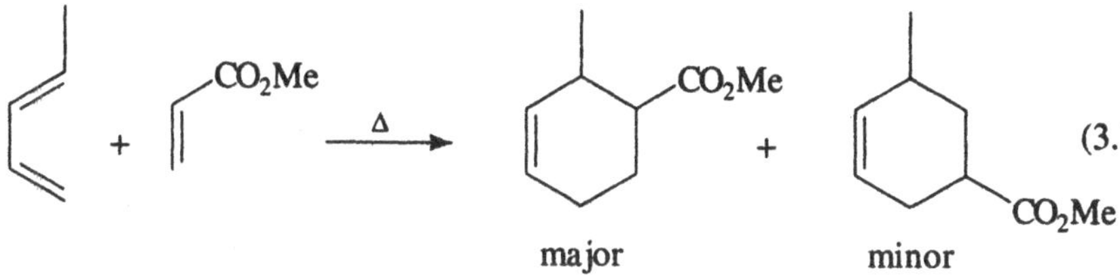

$$\text{major} \qquad \text{minor} \qquad\qquad (3.1)$$

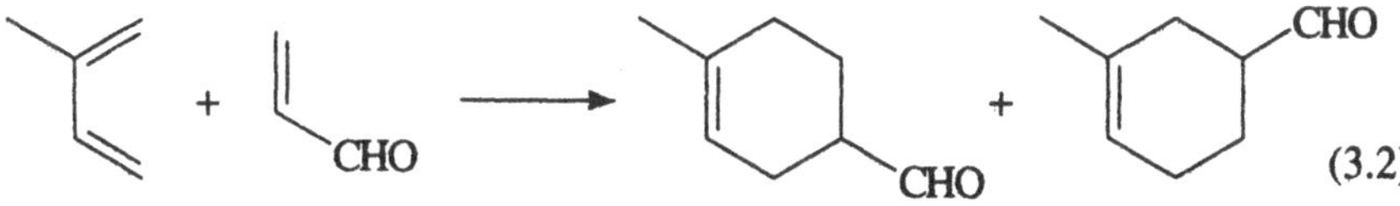

$$(3.2)$$

| toluene, 120°C, no catalyst | 59 | : | 41 |
| benzene, 25°C, SnCl$_4$ | 96 | : | 4 |

As far as the *stereoselectivity* of the reaction is concerned there are two aspects which must be addressed: (i) *exo* versus *endo* selectivity; and (ii) the absolute stereochemical control.

Since the reaction generates up to four new stereogenic centres, 16 possible stereoisomers could in principle be formed. Fortunately, since both components undergo suprafacial addition, the number of possible stereoisomers is reduced (Scheme 3.1).

The *exo* transition state and the *exo* adduct usually involve less steric interactions than the corresponding *endo* transition state and *endo* adduct, but in most cases it is the *endo* adduct which predominates. This is because the *endo*

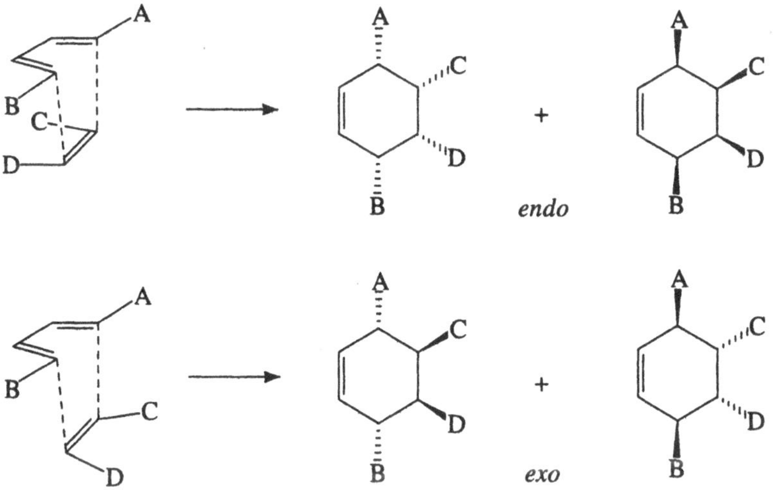

Scheme 3.1

transition state is stabilized by secondary orbital interactions. The effect is to make the less stable *endo* adduct the major product under conditions of kinetic control, when the reaction is effectively irreversible.

When Lewis acid catalysis is employed the stereoselectivity of the reaction is enhanced. The catalyst acts by coordinating to the dienophile, lowering the energy of the LUMO, and hence reducing the energy gap between the HOMO of the diene and the LUMO of the dienophile. The increase in the orbital coefficient on the carbonyl carbon atom also increases the secondary orbital interactions and leads to an increase in the *endo/exo* ratio.

The absolute stereoselectivity of the Diels Alder reaction can be controlled by using either a chiral diene or a chiral dienophile, or by using a chiral Lewis acid. These options are considered in the following sections.

3.1.1 Chiral dienophiles

Many asymmetric Diels Alder reactions depend upon the use of a dienophile containing a chiral auxiliary group. Some examples are shown in eqns 3.3–3.5. The chiral auxiliary used in eqn. 3.3 is derived from camphor, while those used in eqns 3.4 and 3.5 are derived from valine and norephedrine, respectively. In the last two cases the Lewis acid is believed to hold the dienophile in a relatively rigid conformation by coordinating with both of the carbonyl groups (Figure 3.1).

R = H   99% d.e.
R = Me >97% d.e.

>98% *endo*

$$(3.3)$$

R = H   86% d.e.
R = Me   90% d.e.

>98% *endo*

$$(3.4)$$

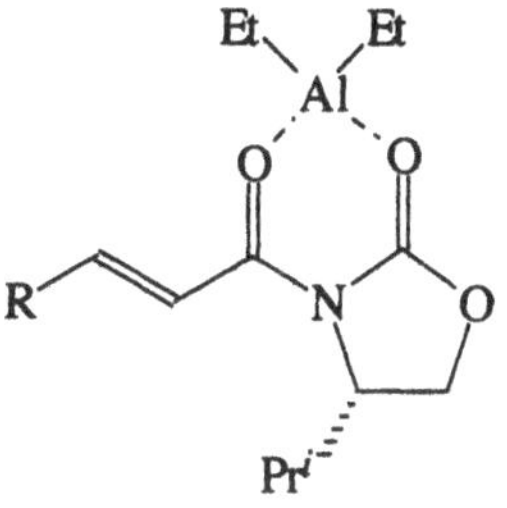

Figure 3.1

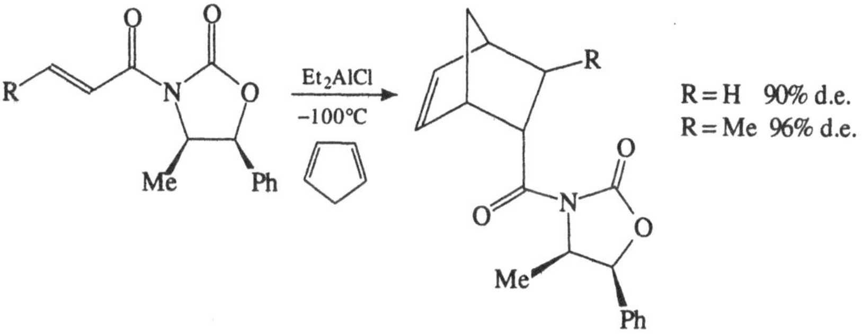

R = H   90% d.e.
R = Me 96% d.e.

(3.5)

3.1.2 Chiral dienes

A number of chiral dienes have been prepared and have been shown to undergo highly stereoselective Diels Alder reactions. One example is shown in eqn. 3.6.

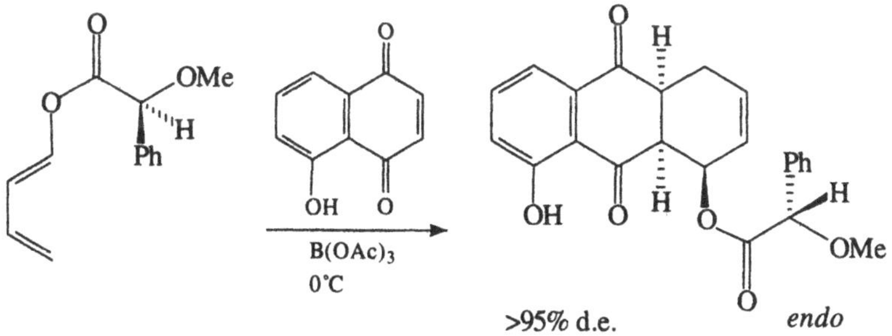

>95% d.e.   *endo*

(3.6)

Winterfeldt *et al.* have used a homochiral diene as a chiral protecting group in order to achieve stereoselective conjugate addition to a cyclohexa-2,5-dienone (Scheme 3.2). As a result of the regioselectivity and stereoselectivity of the cycloaddition only one face of one of the double bonds in the dienone is exposed to attack by the incoming nucleophile. A second example involving the asymmetric synthesis of a butyrolactone from maleic anhydride is shown in Scheme 3.3.

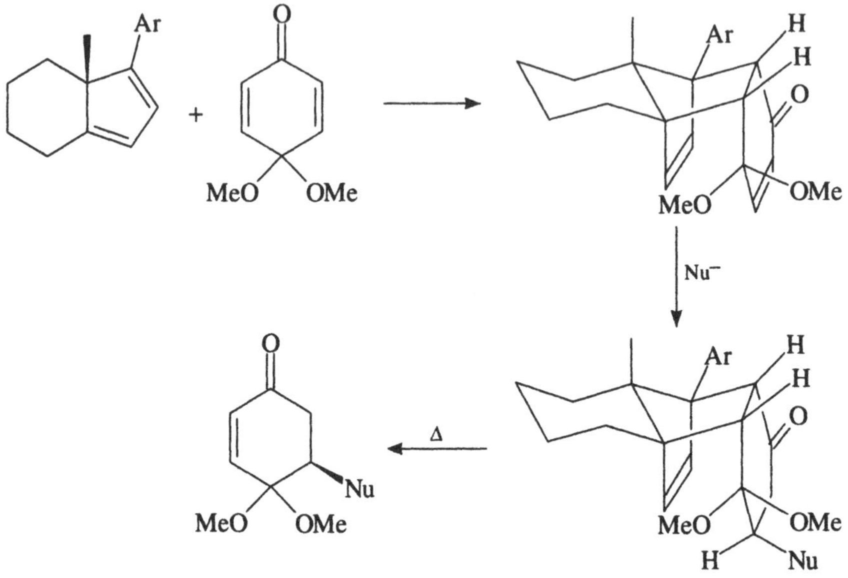

Scheme 3.2

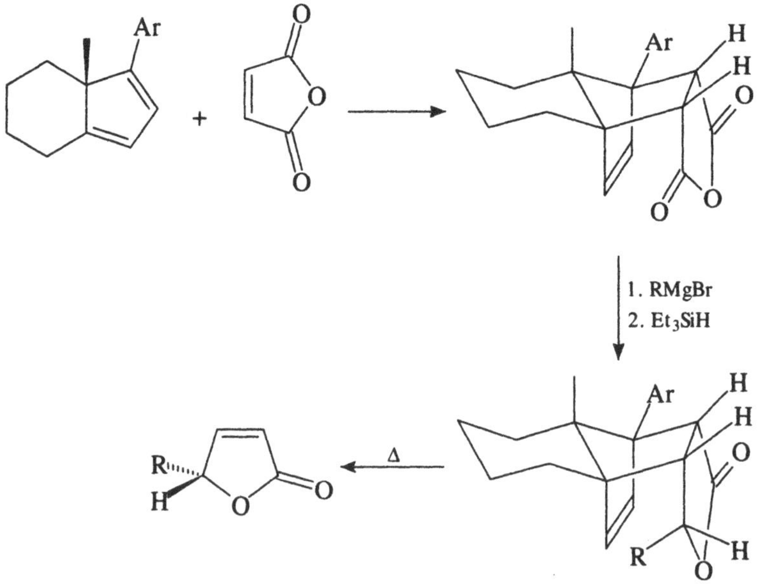

Scheme 3.3

3.1.3 Chiral Lewis acids

Several boron and aluminium based catalysts containing chiral ligands impart very high enantioselectivities on Diels Alder reactions. For example, Corey *et al.* have demonstrated that a diazaaluminolidine is an effective catalyst for the cycloaddition of cyclopentadiene derivatives to activated dienophiles (eqn. 3.7). The ligand is a C_2 symmetric bis(sulfonamide) derivative of 1,2-diamino-1,2-diphenylethane which is formed by reacting the bis(sulfonamide) with $AlMe_3$ *in situ*. Two reactions catalysed by chiral boron containing Lewis acids are shown in eqns 3.8 and 3.9. In the latter case an X-ray analysis of the dienophile-Lewis acid complex supports the schematic representation of the transition state.

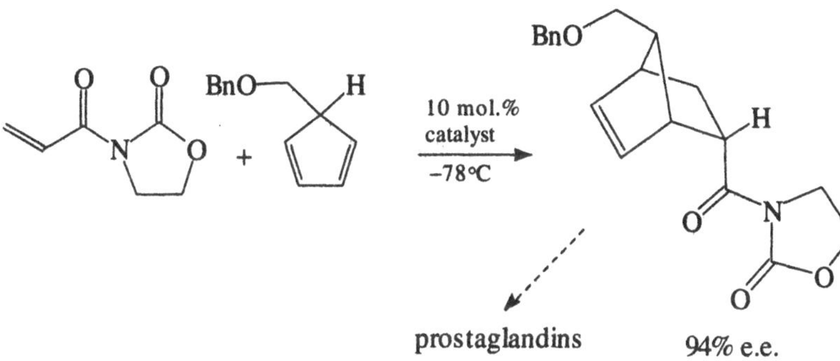

prostaglandins

94% e.e.

(3.7)

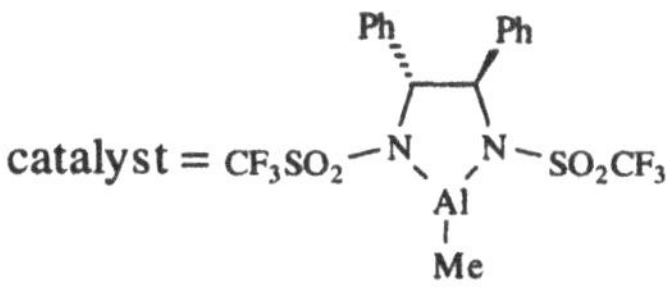

>98% e.e.

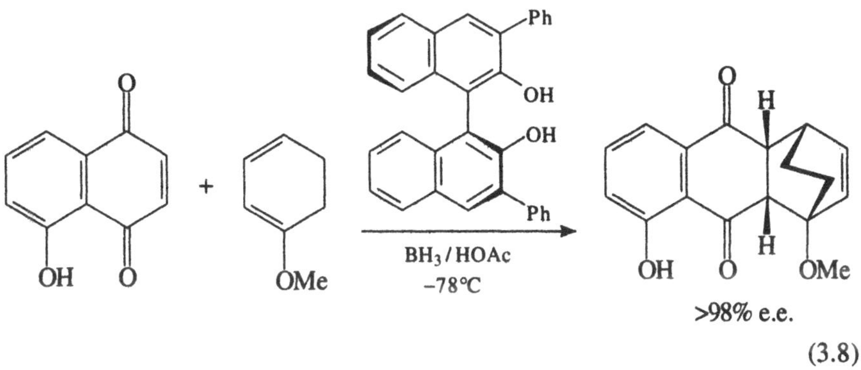

(3.8)

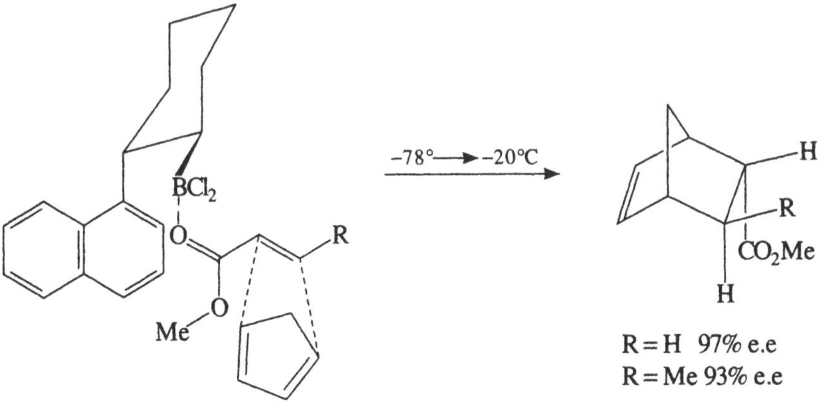

$$-78° \longrightarrow -20°C$$

R = H 97% e.e
R = Me 93% e.e

(3.9)

Corey has also shown that an oxazaborolidine derived from tryptophan imparts extremely high enantioselectivity on the cycloaddition of 3-bromoacrolein to cyclopentadiene. The high selectivity is attributed to an attractive interaction between the dienophile and the indole unit which leads to preferential exposure of one face of the dienophile to the diene (Scheme 3.4). In support of this hypothesis it has been shown that oxazaborolidines derived from other amino acids afford lower selectivities.

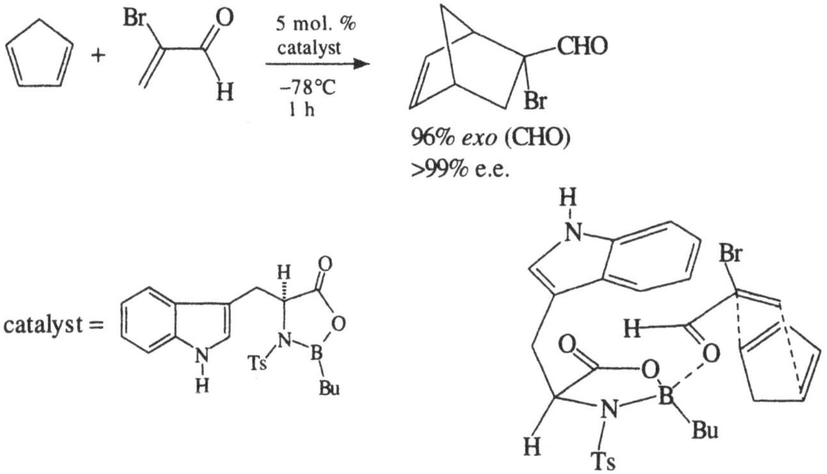

5 mol. %
catalyst
−78°C
1 h

96% *exo* (CHO)
>99% e.e.

catalyst =

Scheme 3.4

3.1.4 Bulky Lewis acids

Yamamoto *et al.* have carried out an elegant study which involves the use of a bulky Lewis acid to enhance the stereoselectivity of the Diels Alder reaction (eqn. 3.10). The reaction of *tert*-butyl methyl fumarate with methyl aluminium *bis*(2,6-di-*tert*-butyl-4-methylphenoxide) (MAD) gives rise to an organoaluminium-fumarate complex (eqn. 3.11), the structure of which was proven by low temperature n.m.r. spectroscopy. The Diels Alder reaction of this complex with cyclopentadiene resulted in the almost exclusive formation of the adduct having the methoxycarbonyl group *endo*. In contrast the use of diethylaluminium chloride as Lewis acid was found to have a total lack of selectivity. Even ethyl and methyl esters can be effectively differentiated by MAD (71 : 29).

heat / 80°C	48	:	52
Et_2AlCl / $-78°C$	46	:	54
MAD / $-78°C$	99	:	1

(3.10)

MAD

(3.11)

The asymmetric Diels Alder reaction of (−)-menthyl methyl fumarate with cyclopentadiene in dichloromethane under the influence of MAD proceeds in 86% d.e. with an *endo/exo*-(CO_2Me) ratio of 98·4 : 1·6. In contrast, the diethylaluminium chloride catalysed cycloaddition gave an 80% d.e. and an *endo/exo* ratio of only 57:43 (eqn. 3.12).

heat / 80°C	22·8	:	26·4	:	24·0	:	26·8
Et₂AlCl / −78°C	52·5	:	4·6	:	5·2	:	37·7
MAD / −78°C	91·4	:	7·0	:	0·2	:	1·4

$$(3.12)$$

3.2 [2 + 2] CYCLOADDITION REACTIONS

Many [2 + 2] cycloaddition reactions proceed in a concerted manner, while others involve a stepwise mechanism involving ionic or radical intermediates. The concerted reactions in particular are highly stereoselective and enantiomerically enriched products can be prepared by incorporating a chiral auxiliary in one of the components (eqns 3.13 and 3.14).

86% d.e.   2 : 1

$$(3.13)$$

$$(3.14)$$

R* = (−)-8-phenylmenthyl   R = Me   20% d.e.
                            R = Bu^i 60% d.e.
                            R = Bu^t 96% d.e.
                            R = Ph   96% d.e.

3.3 SIGMATROPIC REARRANGEMENTS

Another group of pericyclic reactions whose outcome is dictated by the geometrical requirements of a cyclic transition state are sigmatropic rearrangements. Some examples of the control operative in [3,3] sigmatropic rearrangements are illustrated below. The observed stereoselectivity can usually be rationalized by assuming a preference for a chair-like transition state in which 1,3-diaxial interactions are minimized. Scheme 3.5 shows an example of a Cope rearrangement involving a 1,5-diene containing a stereogenic centre. The requirement for the larger phenyl group to adopt an equatorial configuration in the transition state effectively dictates both the configuration of the new stereogenic centre and the trisubstituted double bond. Equations 3.15–3.17 show variants of the well known Claisen rearrangement in which the same considerations apply.

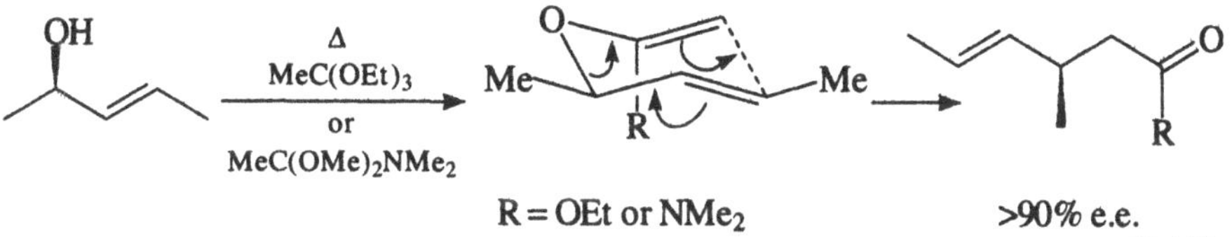

$$R = OEt \text{ or } NMe_2 \qquad >90\% \text{ e.e.}$$

$$(3.15)$$

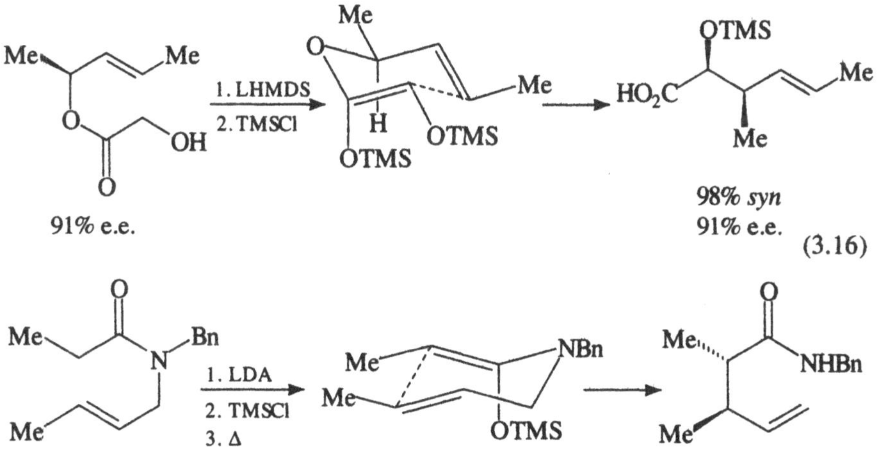

98% syn
91% e.e.

91% e.e.

$$(3.16)$$

99% d.e.

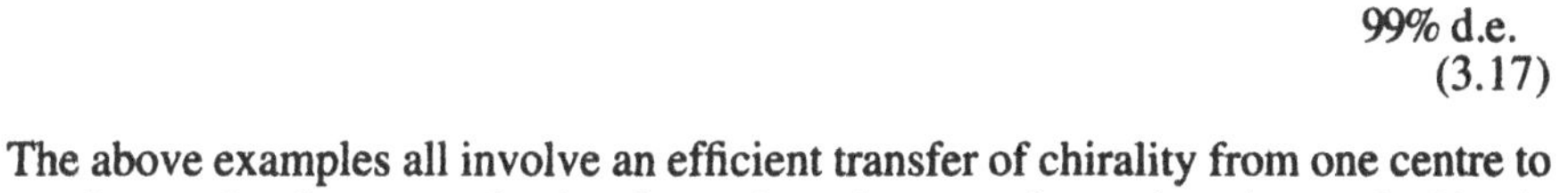

$$(3.17)$$

The above examples all involve an efficient transfer of chirality from one centre to another or the diastereoselective formation of a racemic product due to the highly ordered transition state of the reaction. More impressive are examples in which new stereogenic centres are created with the aid of a chiral auxiliary or a chiral catalyst (eqns 3.18–21). In each case the appropriate chair-like transition state is shown but no attempt is made to depict the role of the chiral auxiliary or catalyst.

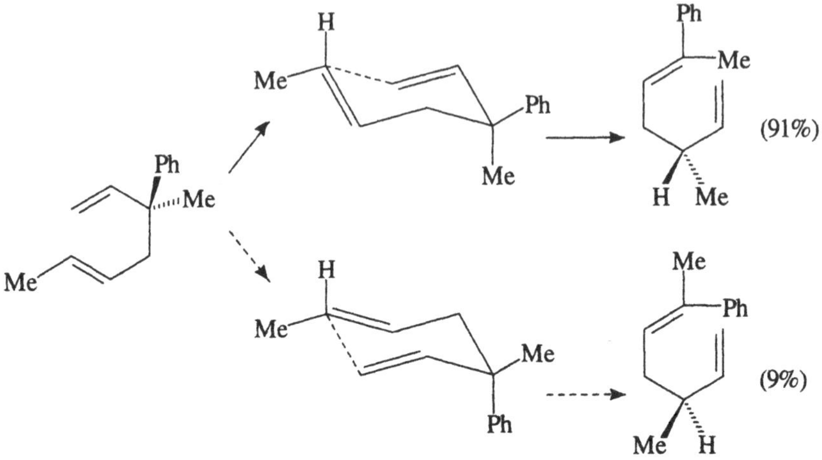

Scheme 3.5

Readers are referred to the suggested further reading for greater discussion of this aspect.

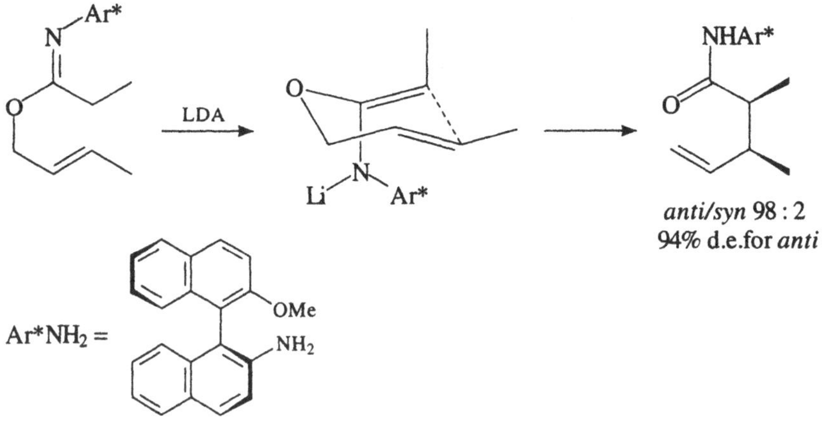

anti/syn 98 : 2
94% d.e. for *anti*

$$Ar^*NH_2 =$$

(3.18)

96% e.e.

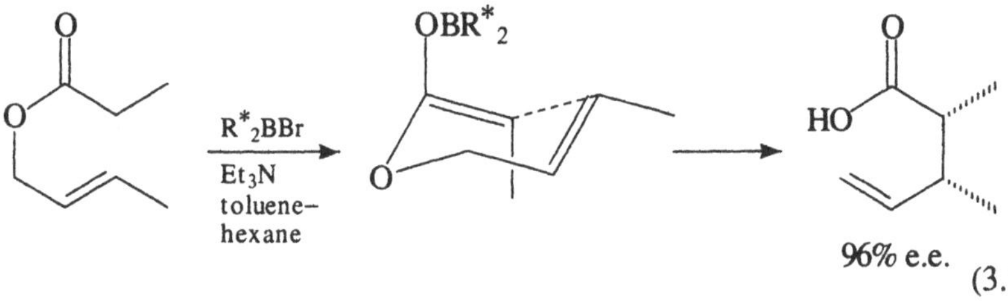

(3.19)

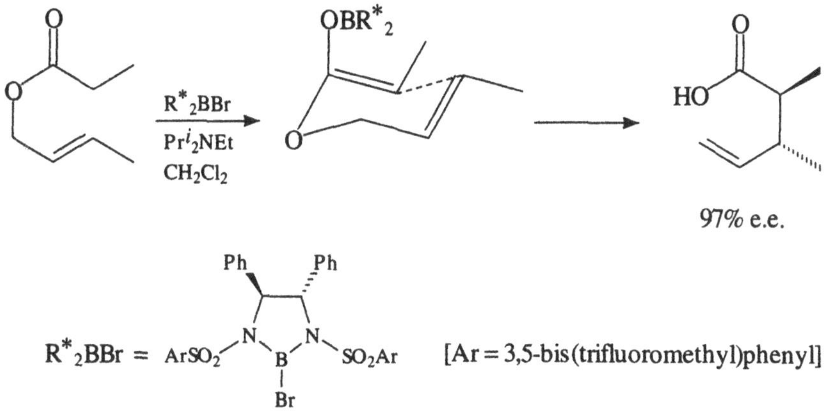

(3.20)

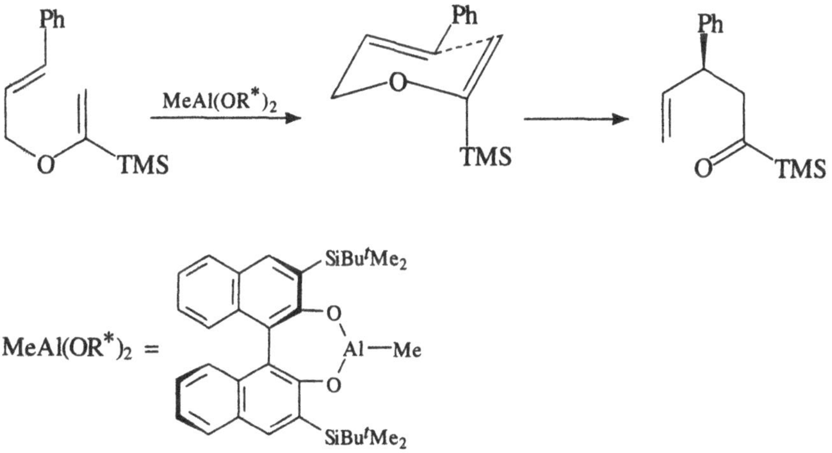

(3.21)

PROBLEMS

1. Give the structure and stereochemistry of the product of the following reaction:

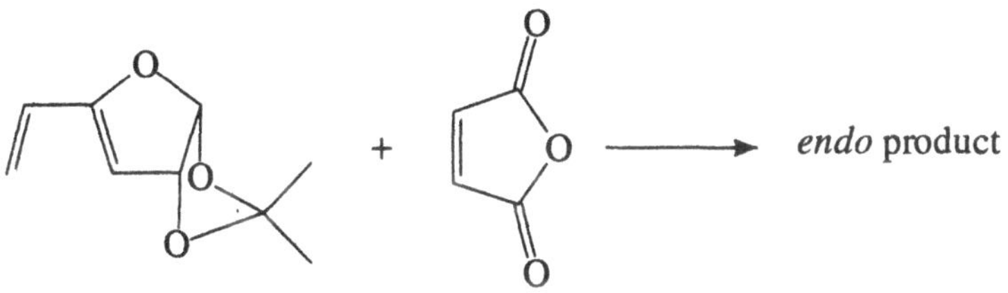

+ $\longrightarrow$ *endo* product

2. Draw the structures of the four possible stereoisomeric products (two *endo* and two *exo*) of the following reaction:

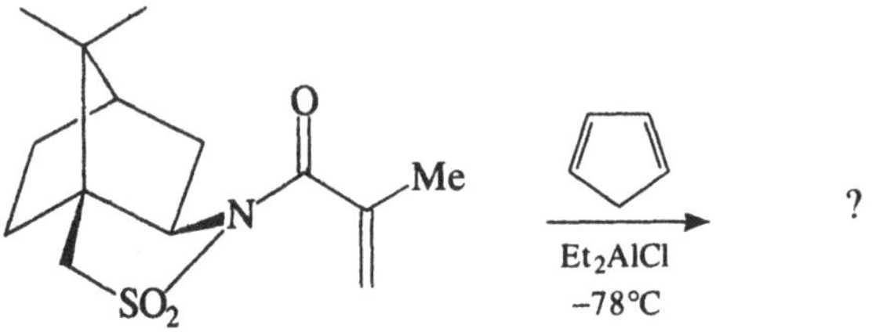
O
Me
N
SO2
Et2AlCl
–78°C
?

3. Rationalize the following transformations:

(a)

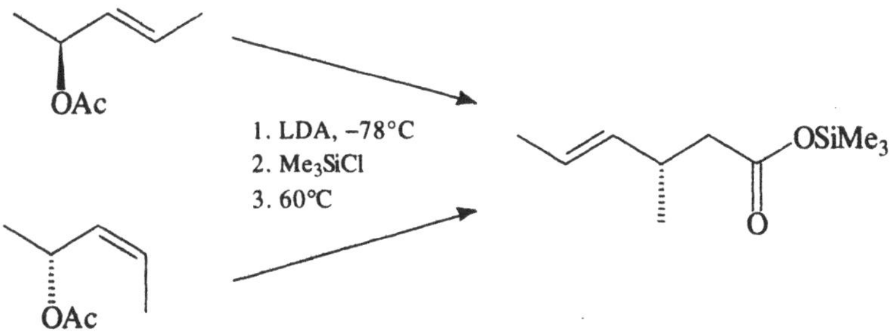
OAc
1. LDA, –78°C
2. Me3SiCl
3. 60°C
OSiMe3
O
OAc

(b)

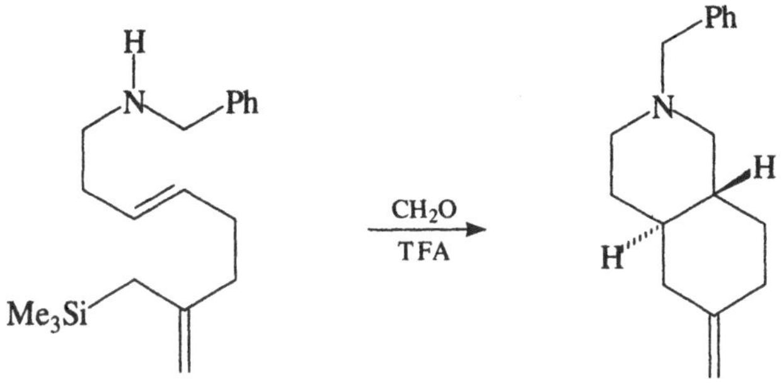
H
N
Ph
Me3Si
CH2O
TFA
Ph
N
H
H

(c)

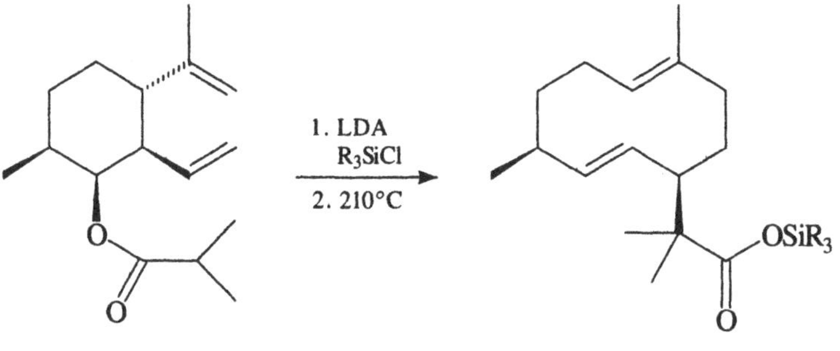
O
O
1. LDA
R3SiCl
2. 210°C
OSiR3
O

FURTHER READING

W. Carruthers, *Cycloaddition Reactions in Organic Chemistry*, Pergamon Press, Oxford, 1990, chaps 1–3 and 7.

Chiral dienophiles

W. Oppolzer, *Angew. Chem. Int. Ed.*, 1984, *23*, 875.
W. Oppolzer, C. Chapuis and G. Bernardinelli, *Helv. Chim. Acta*, 1984, *67*, 1397.
D. A. Evans, K. T. Chapman, D. T. Hung and A. T. Kawaguchi, *Angew. Chem. Int. Ed.*, 1987, 26, 1184.
D. A. Evans, K. T. Chapman and J. Bisaha, *J. Am. Chem. Soc.*, 1988, *110*, 1238.

Chiral dienes

B. M. Trost, S. A. Godelski and J. P. Genet, *J. Am. Chem. Soc.*, 1978, *100*, 3930.
B. M. Trost, D. O'Krongly and J. L. Belletire, *J. Am. Chem. Soc.*, 1980, *102*, 7595.
C. Borm, D. Meibom and E. Winterfeldt, *J. Chem. Soc., Chem. Commun.*, 1996, 887.
E. Winterfeldt, *Chem. Rev.*, 1993, *93*, 827.

Chiral Lewis acids

E. J. Corey, N. Imai and S. Pikul, *Tetrahedron Lett.*, 1991, *32*, 7517.
E. J. Corey and S. Sarshar, *J. Am. Chem. Soc.*, 1992, *114*, 7938.
J. M. Hawkins and S. Loren, *J. Am. Chem. Soc.*, 1991, *113*, 7794.
E. J. Corey and T. P. Loh, *J. Am. Chem. Soc.*, 1991, *113*, 8966.
D. Sartor, J. Saffrich, G. Helmchen and C. J. Richards, *Tetrahedron Asymmetry*, 1991, 2, 639.
K. Mikami, M. Terada, Y. Motoyama and T. Nakei, *Tetrahedron Asymmetry*, 1991, 2, 643.

Bulky Lewis acids

K. Maruoka, S. Saito and H. Yamamoto, *J. Am. Chem. Soc.*, 1992, *114*, 1089.
K. Maruoka, M. Akakura, S. Saito and H. Yamamoto, *J. Am. Chem. Soc.*, 1994, *116*, 6153.
S. Saito and H. Yamamoto, *J. Chem. Soc., Chem. Commun.*, 1997, 1585.

Claisen rearrangements

P. Metz and B. Hungerhoff, *J. Org. Chem.*, 1997, *62*, 4442.
E. J. Corey and D. H. Lee, *J. Am. Chem. Soc.*, 1991, *113*, 4026.
K. Maruoka, H. Banno and H. Yamamoto, *J. Am. Chem. Soc.*, 1990, *112*, 7791.
K. Maruoka, H. Banno and H. Yamamoto, *Tetrahedron Asymm.*, 1991, 2, 647.
K. Maruoka and H. Yamamoto, *Synlett.*, 1991, 793.

ANSWERS

1.

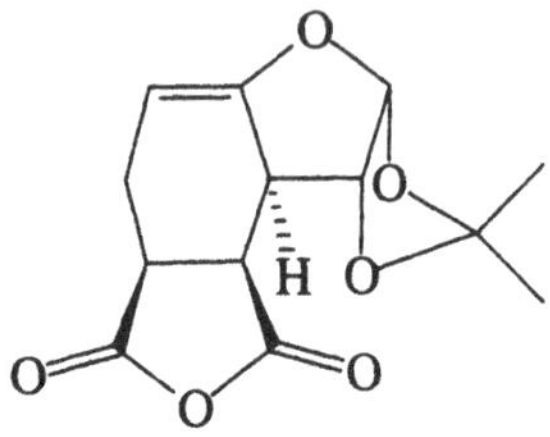

2.

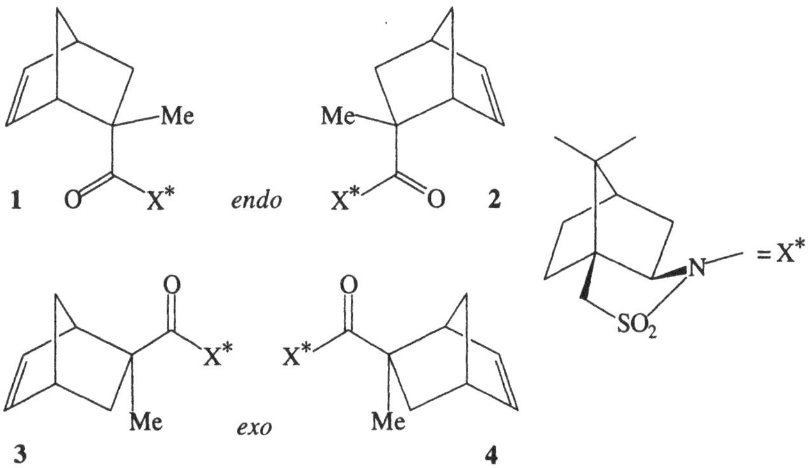

1   O X*   *endo*   X*   O   **2**

= X*

3   Me   *exo*   Me   **4**

In practice the *exo* adducts predominate (73:27). The four adducts (**1–4**) were obtained in the ratio 26:1·4:32·5:40·1. Hence the d.e. of the *endo* adduct is 90% while that of the *exo* adduct is just 10% (Oppolzer, B. M. Seletsky and G. Bernardinelli, *Tetrahedron Lett.*, 1994, *35*, 3509).

3. (a)

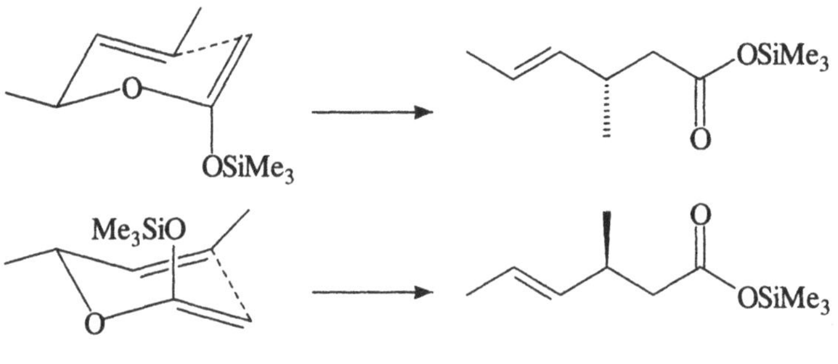

(b)

(c)

4

Stereoselective Reactions of Carbonyl Compounds II

In this chapter we consider three related aspects of the reactivity of carbonyl compounds: (a) enolate formation; (b) reactions of enolates with electrophiles; and (c) the aldol reaction (Scheme 4.1).

For unsymmetrically substituted ketones the *regioselectivity* of reactions (b) and (c) depend upon the regioselectivity of enolate formation (step a). The regioselectivity of this step can be studied by trapping the enolates as their silyl

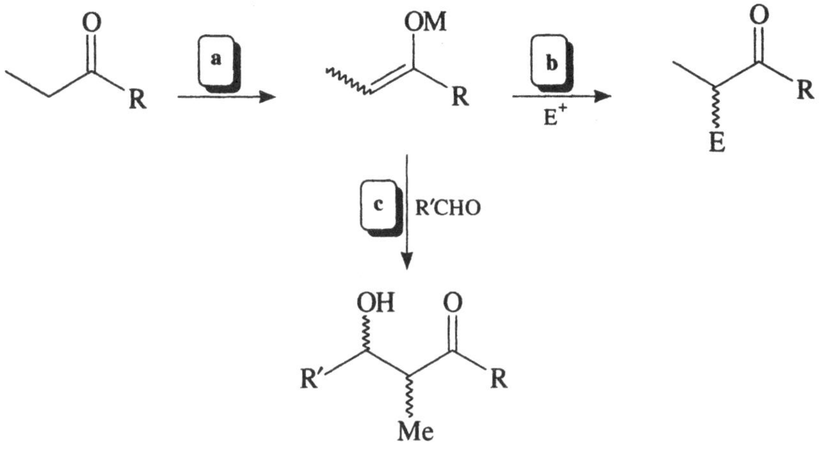

Scheme 4.1

enol ethers by reaction with trimethylsilyl chloride (which reacts exclusively on oxygen due to the strength of the Si–O bond). The relative proportions of the two silyl enol ethers formed depends upon the conditions under which the enolates are generated.

For example, heating 2-methylcyclohexanone with triethylamine gives mainly the so-called "thermodynamic" enolate since the silyl enol ether (1) predominates (Scheme 4.2). Under these conditions (weak base, high temperature) the two enolates equilibrate and the one which is thermodynamically more stable (because it contains a more highly substituted double bond) predominates. The composition of the mixture of silyl enol ethers therefore reflects the relative stability of the two enolates.

In contrast, when 2-methylcyclohexanone is treated with lithium diisopropylamide (a strong base) at low temperature (no equilibration) the alternative silyl enol ether (2) predominates. These conditions do not allow equilibration between the two enolates and hence the one which predominates is the one which is formed fastest rather than the one having the greater stability. This is described as the "kinetic" enolate and it is formed more rapidly because a proton can be abstracted more easily from the less hindered α-carbon atom.

The regioselectivity of enolate formation determines the regioselectivity of subsequent reactions (eqns 4.1 and 4.2).

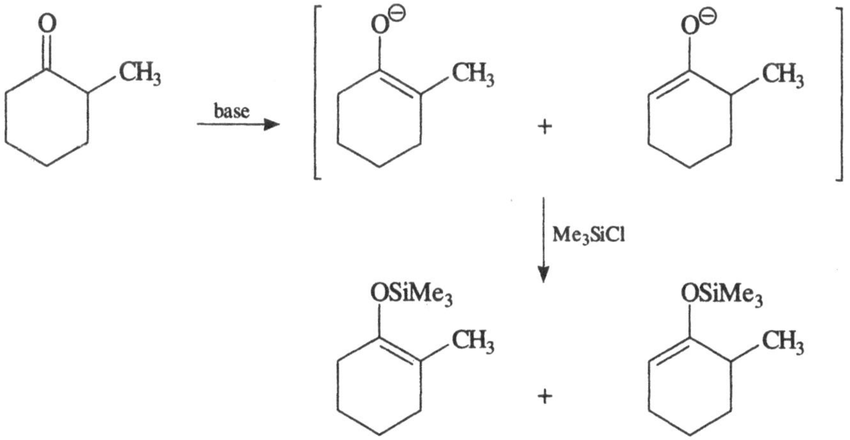

	1		2
Et₃N + heat	78	:	22
LDA < 0° C	1	:	99

Scheme 4.2

(4.1)

(4.2)

4.1 STEREOSELECTIVITY OF ENOLATE FORMATION

The stereochemical options are illustrated in Scheme 4.3. The stereoselectivity of reactions (b) and (c) depend upon the stereoselectivity of step (a), as well as on other factors. Note that the convention adopted in naming the aldol products is that when the carbon chain is written in an extended (zigzag) form, the *syn* isomers have the OH and Me groups on the same side (i.e. both above or both below the plane of the paper) and the *anti* isomers have them on opposite sides (i.e. one above and one below).

For most of the reactions considered in this chapter the conditions used or the compound itself dictate that equilibration of the enolates is not an option. Lithium enolates will be considered in the first instance since they have been extensively investigated. However it will be clear from later examples that many other elements can be utilized and that boron enolates, for example, have advantages in some situations. The relative proportions of the Z and E enolates vary depending upon the nature of the group R (eqn. 4.3).

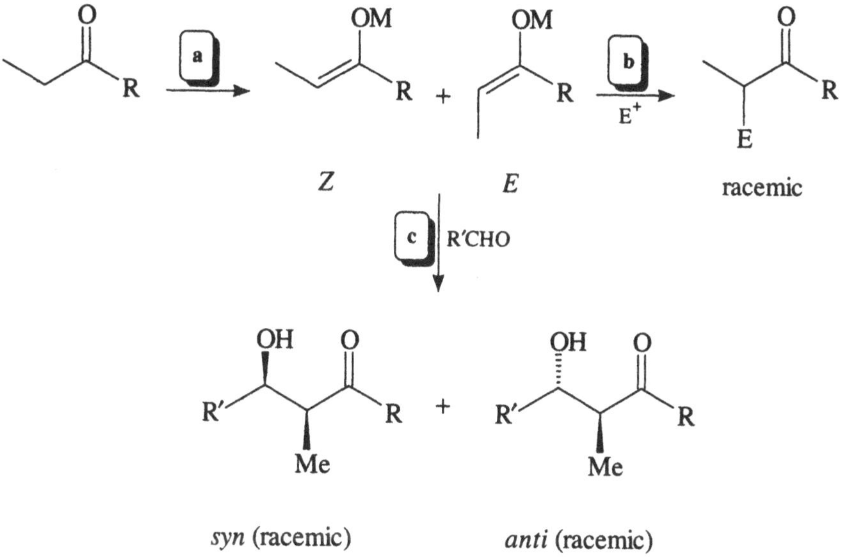

Scheme 4.3

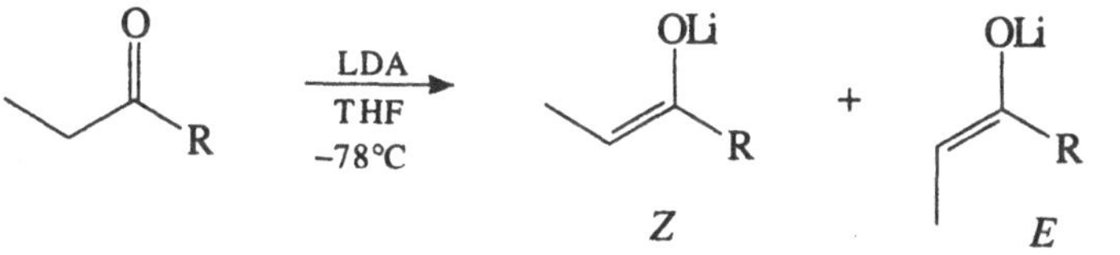

R = Et	30	:	70
Pr^i	60	:	40
Bu^t	>98	:	<2
OMe	5 (E)	:	95 (Z)
NEt$_2$	>97	:	<3

$$(4.3)$$

It can be seen that for simple ketones the *E* enolate is formed when R is small, but the *Z* enolate is formed if R is large. This can be rationalized by considering the two conformations of the parent ketone from which proton abstraction occurs (Figure 4.1). When R is small the transition state leading to the *Z* enolate is destabilized by the 1,3-diaxial interaction between the methyl group and the *iso*-propyl group. When R is large the conformation leading to the *E* enolate is destabilized by steric interaction between the methyl group and the R group. When R is an alkoxy group this interaction is alleviated with the result that the *E* enolate (actually *Z* in this case due to the priority of OMe over OLi) is preferred (Figure 4.2). In the amide the *Z* enolate is once again preferred (Figure 4.3).

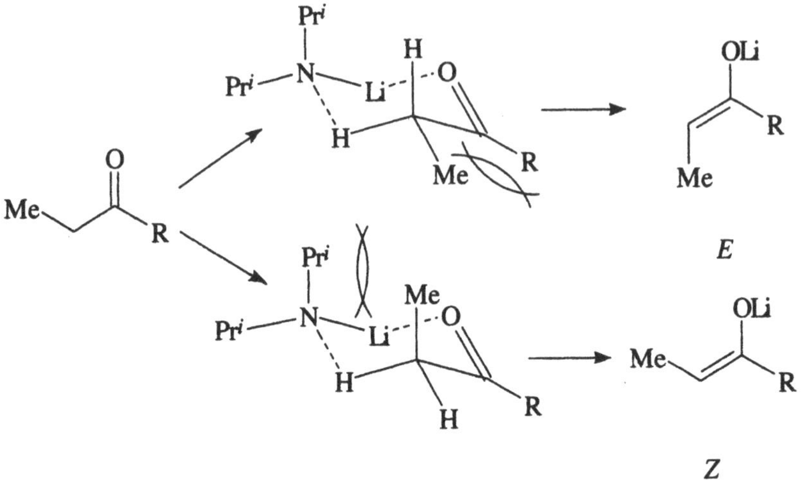

Figure 4.1

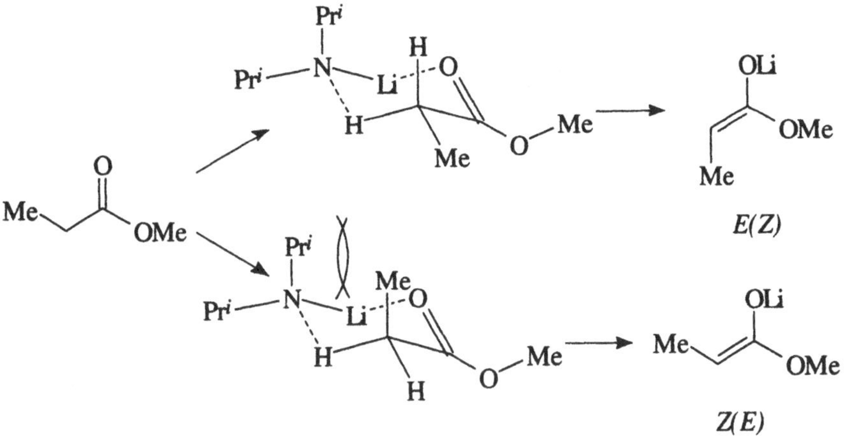

Figure 4.2

However the Z/E ratio depends upon many other factors besides the nature of the substituent groups, and can be influenced in a number of ways (eqns 4.4 and 4.5). For example, the use of additives such as hexamethylphosphoric triamide (HMPA) reduces the coordinating power of the cation and leads to equilibration to give the thermodynamically preferred Z enolate.

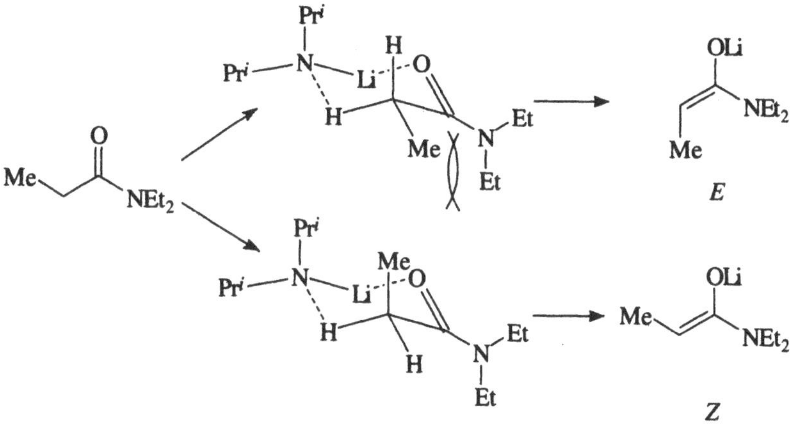

Figure 4.3

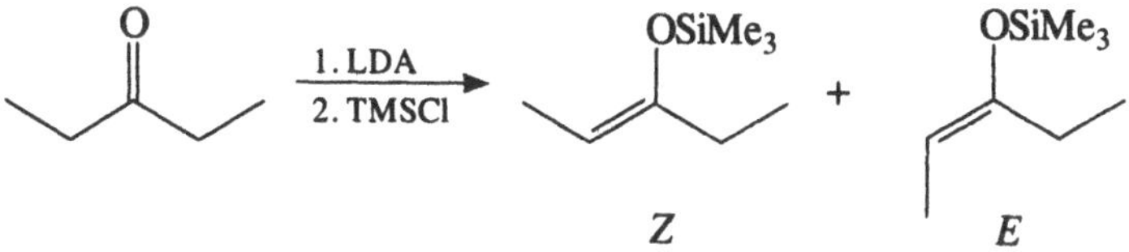

THF	30	:	70
THF/HMPA	82	:	18

HMPA = (Me$_2$N)$_3$PO

(4.4)

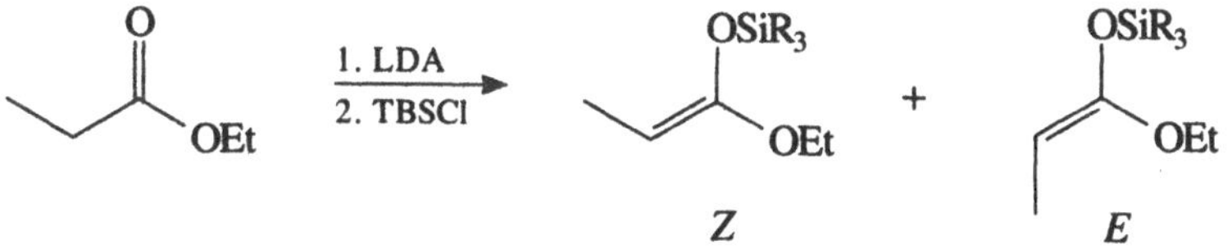

THF	6	:	94
THF/HMPA	82	:	18

(4.5)

4.2 ALKYLATION OF ENOLATES

Whichever enolate is formed it will undergo alkylation to give a racemic product unless a chiral influence is present. Diastereoselective alkylation can, for example, be achieved by incorporating a chiral auxiliary group. Evans *et al.* have introduced auxiliary groups derived from valine and norephedrine which lead to either isomer of the α-alkylated product (eqns 4.6 and 4.7).

$$R = CH_2=CHCH_2 \quad 94\% \text{ d.e.}$$
$$R = Et \qquad\qquad 90\% \text{ d.e.}$$

valine

(4.6)

$$R = CH_2=CHCH_2 \quad 94\% \text{ d.e.}$$
$$R = Et \qquad\qquad 90\% \text{ d.e.}$$

norephedrine

(4.7)

An alternative strategy introduced by Meyers *et al.* involves a chiral aza-enolate of an oxazoline prepared by reacting a carboxylic acid with an aminoalcohol (Scheme 4.4). The methoxymethyl group plays a crucial role in coordinating the lithium cation which is involved in the proton abstraction step and in the delivery of the alkyl group.

Enders *et al.* introduced the chiral hydrazines SAMP and RAMP (the two enantiomers of 1-amino-2-methoxymethylpyrrolidine) to enable the stereoselective alkylation of ketones (Scheme 4.5). The methoxymethyl group again plays a crucial role in determining the site of the proton abstraction and in assisting in the delivery of the alkyl halide.

Myers *et al.* have reported an "exceedingly practical" method for the asymmetric synthesis of α-amino acids employing as a key step the alkylation of pseudoephedrine glycinamide, which is readily prepared from pseudoephedrine and glycine methyl ester. The alkylation reaction proceeds in high yield and the products are formed with high diastereoselectivity (Scheme 4.6). Almost two equivalents of LDA are required which, in the presence of anhydrous lithium chloride, lead to kinetic deprotonation of the hydroxyl and amino groups to

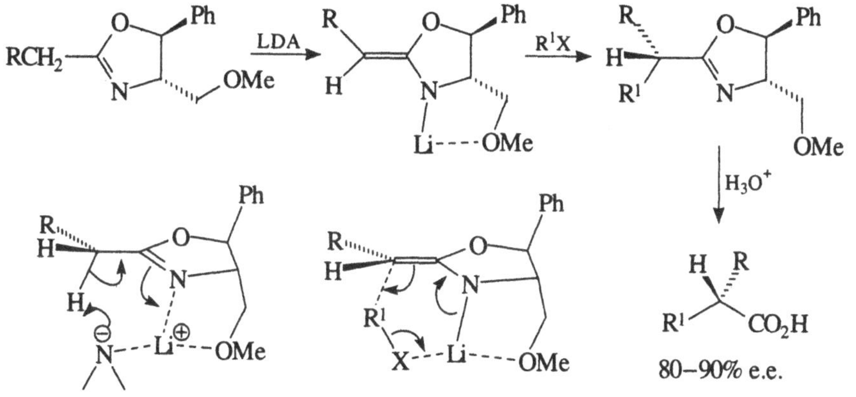

Scheme 4.4

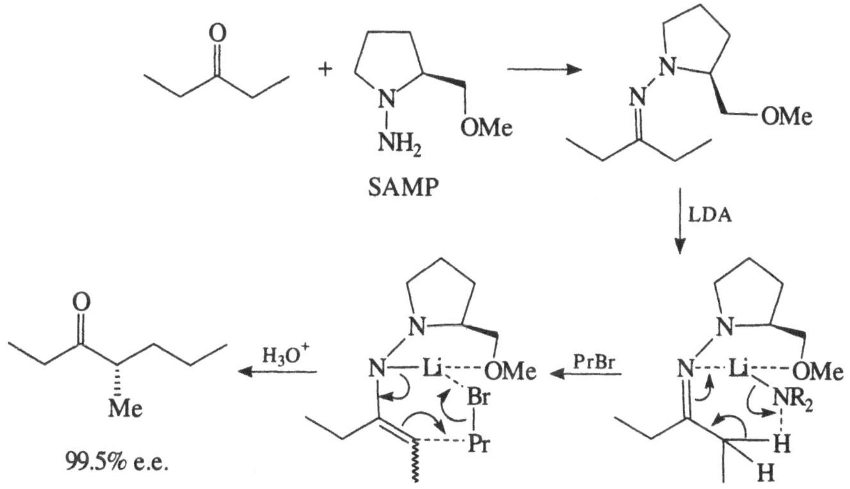

Scheme 4.5

generate an *O,N*-dianion. Warming the reaction mixture to 0°C leads to equilibration to the thermodynamically more stable enolate anion, which reacts with the alkyl halide to afford a *C*-alkylated product. This can be hydrolysed efficiently, with little or no racemization, simply by heating with aqueous sodium hydroxide to give the α-amino acid or its *N*-protected derivative. A working model to account for the diastereoselectivity of the alkylation step invokes the blocking of the enolate π face by the secondary lithium alkoxide and, perhaps more importantly, the solvent (THF) molecules associated with the lithium cation.

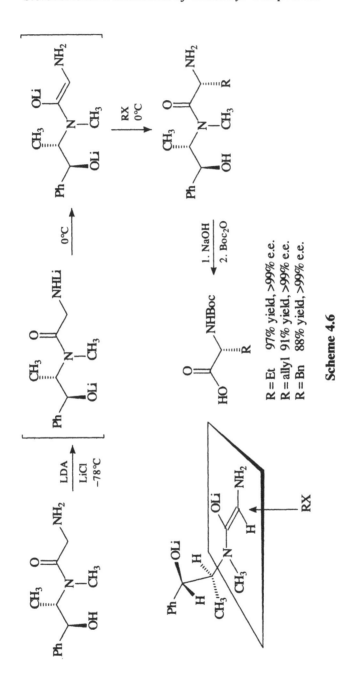

Scheme 4.6

Other electrophiles besides alkyl groups can also be introduced stereoselectively into chiral enolates leading, for example, to α-hydroxy and α-amino acids (eqns 4.8–4.10).

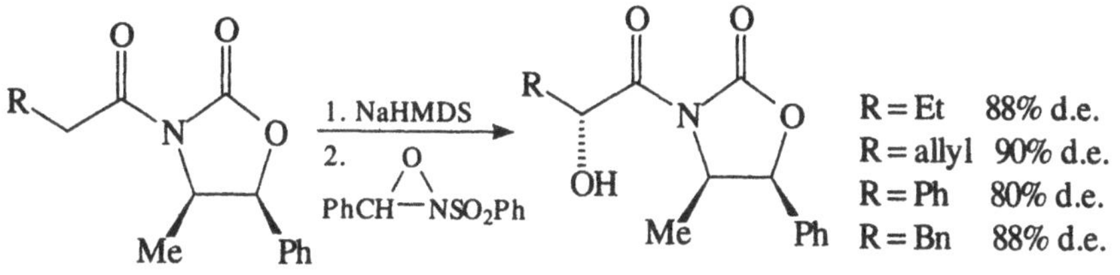

$$R = Et \quad 88\% \text{ d.e.}$$
$$R = \text{allyl} \quad 90\% \text{ d.e.}$$
$$R = Ph \quad 80\% \text{ d.e.}$$
$$R = Bn \quad 88\% \text{ d.e.}$$

(4.8)

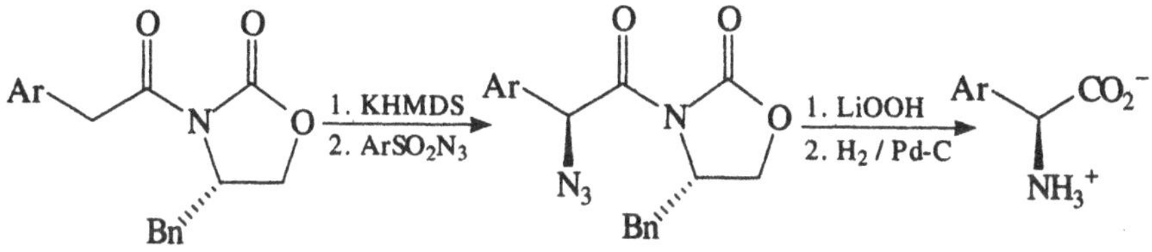

(4.9)

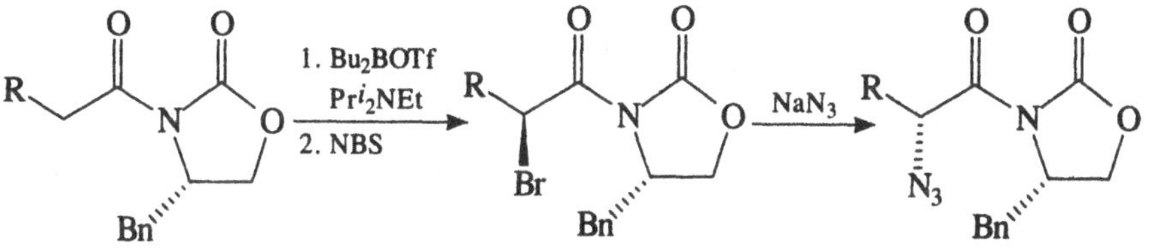

(4.10)

4.3 THE ALDOL REACTION

In the aldol reaction two new stereogenic centres are formed and two diastereoisomers are therefore produced (eqn. 4.11).

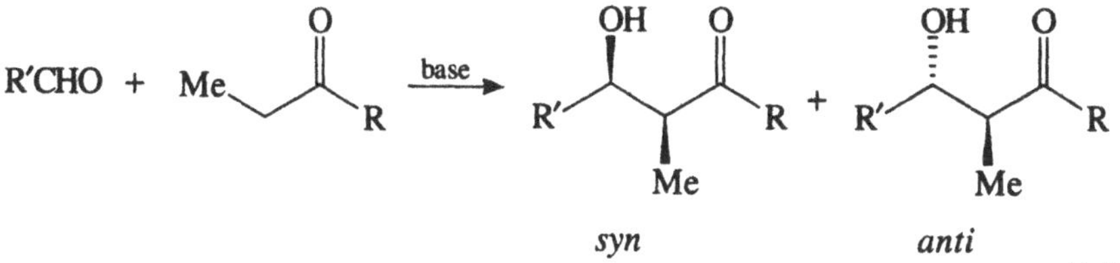

(4.11)

In contrast to the addition of allyl boron derivatives (Section 2.3) which is irreversible and thus under kinetic control, the aldol reaction is reversible and may be effected under either kinetic or thermodynamic control. Under kinetic control,

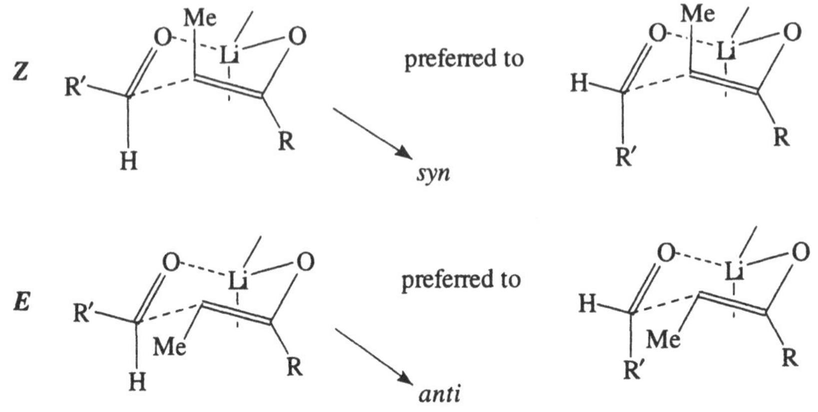

Figure 4.4

in the majority of cases, a cyclic transition state is involved in which the metal atom is coordinated to the carbonyl oxygen of the other component. The major diastereomer depends upon the enolate involved and can be predicted using the so-called Zimmerman–Traxler model (Figure 4.4). This assumes that the six-membered transition state adopts a chair shape and that the usual principles of conformational analysis apply. Of the two possible transition states involving the Z enolate the one which places the R′ group of the aldehyde equatorial is preferred, leading to the *syn* product. Applying the same analysis to the E enolate leads to the prediction that the *anti* aldol will be preferred.

In order to fully appreciate the following results it is necessary to make two further assumptions: (i) that Z enolates are more stereoselective than E enolates; and (ii) that boron enolates are more stereoselective than Li enolates.

The latter assumption can be understood on the basis that the B–O bond is shorter than the Li–O bond so that the transition state involving B is much tighter and steric effects are magnified, leading to greater selectivity (eqns 4.12 and 4.13).

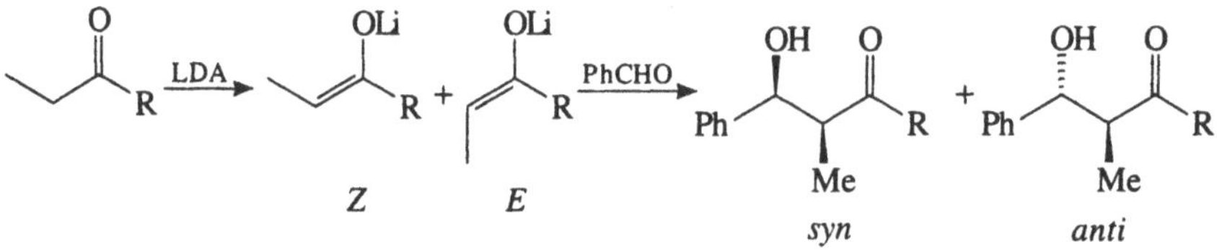

	Z	:	E		*syn*	:	*anti*
R = Et	30	:	70		64	:	36
Pr^i	60	:	40		82	:	18
Bu^t	>98	:	<2		>98	:	<2

$$(4.12)$$

$$Z \ + \ E \ \xrightarrow{\text{PhCHO}} \ syn \ + \ anti$$

$$30 \ : \ 70 \qquad\qquad 64 \ : \ 36$$

$$\xrightarrow[\text{Pr}^i\text{NEt}]{\text{Bu}_2\text{BOTf}} \quad 98 \ : \ 2 \qquad\qquad 98 \ : \ 2$$

(4.13)

Similarly, although esters give *E*-enolates with LDA the *syn/anti* selectivity is poor, but it can be dramatically improved by the use of the boron enolate (eqn 4.14).

$$\xrightarrow{\text{PhCHO}} \quad syn \ + \ anti$$
$$49 \ : \ 51$$

$$E(Z)$$

$$\xrightarrow[\substack{\text{Et}_3\text{N} \\ \text{Toluene}}]{\text{Bu}_2\text{BOTf}} \quad E(Z) \quad \longrightarrow \quad anti \ \text{only}$$

(4.14)

For enolates giving the same major product irrespective of their configuration, or for reactions requiring Lewis acid catalysis, either thermodynamic control or an open-chain transition state should be suspected (see later).

4.4 ASYMMETRIC ALDOL REACTIONS

Incorporating a chiral auxiliary into either the aldehyde or the (*Z*)-enolate leads to a predominance of one of the two diastereomeric *syn* aldols. When both components contain a chiral group then they will both influence the stereoselectivity of the reaction. This phenomenon is described as *double asymmetric induction*. The two groups will either both favour the formation of the same isomer (a matched pair) or their effects will oppose one another (a mismatched pair). Both of these possibilities are illustrated in Scheme 4.7. In the case of the matched pair a high diastereomeric excess can be obtained. The preferred stereoselectivity of the chiral enolate can be partly understood by considering the conformations of the two cyclic transition states (Figure 4.5). Molecular models would be needed to fully appreciate the difference in stability between the two possible arrangements.

By employing boron enolates having extremely high diastereofacial selectivities Evans *et al.* have developed reagents which can completely override the stereochemical preference of the aldehyde (Scheme 4.8). The preferred stereochemical course of the reaction involving the valine derived auxiliary is illustrated in Scheme 4.9.

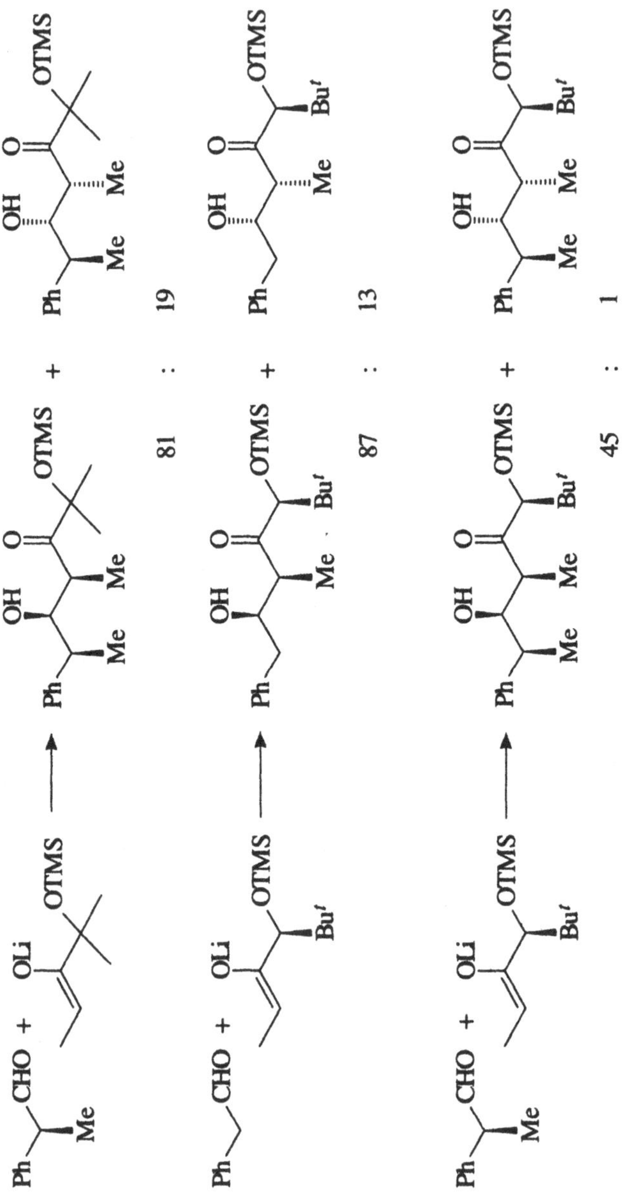

Scheme 4.7

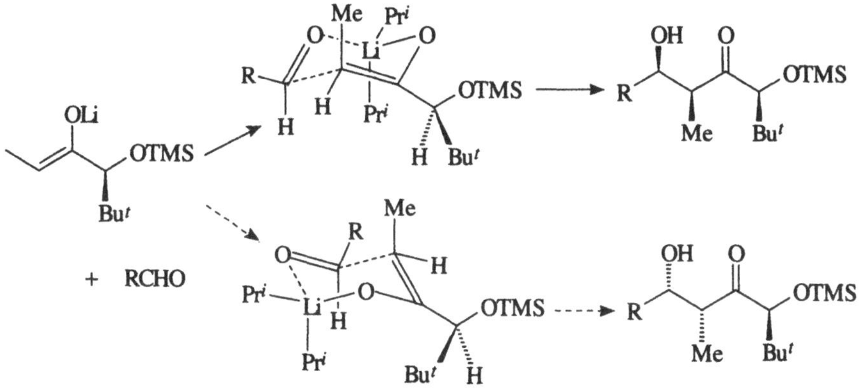

Figure 4.5

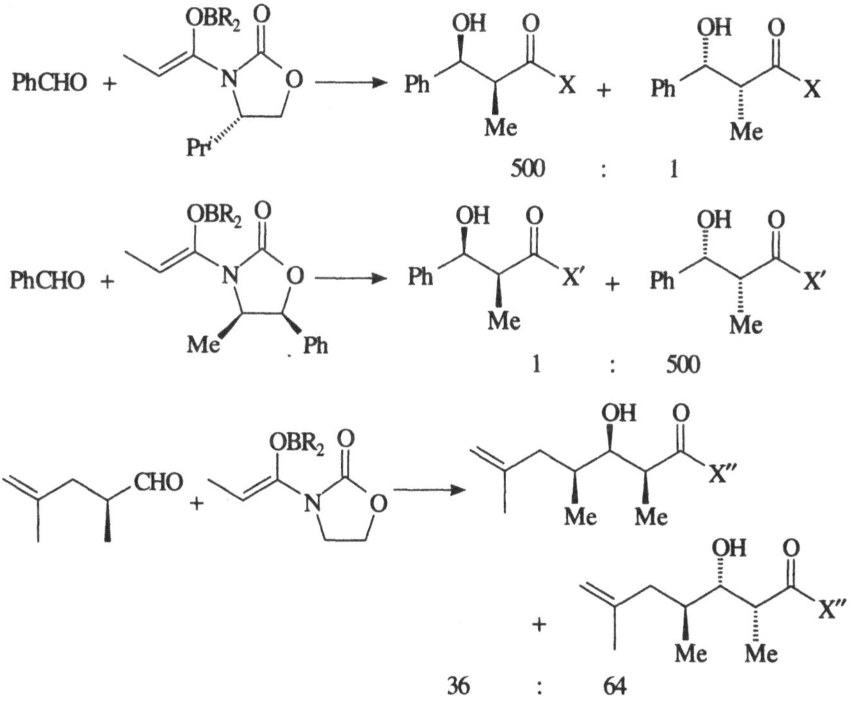

Scheme 4.8

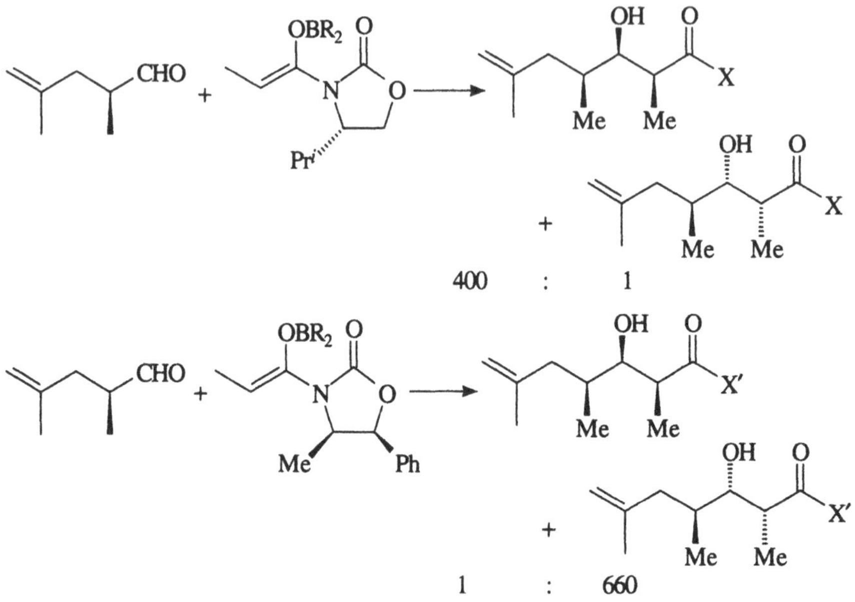

Scheme 4.8 *(continued)*

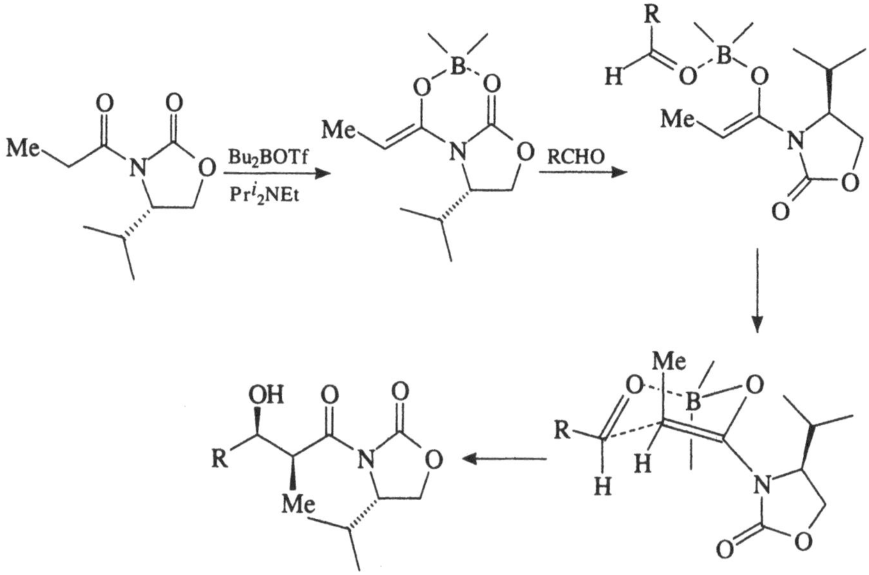

Scheme 4.9

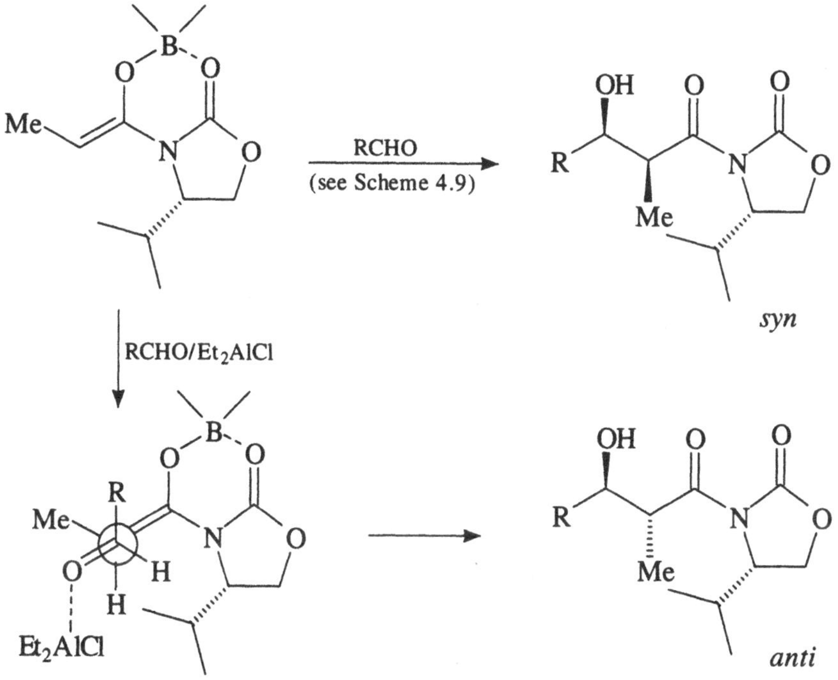

Scheme 4.10

The *syn/anti* ratio can be altered by using Lewis acid catalysts which favour the formation of an open (non-cyclic) transition state since the aldehyde coordinates with the Lewis acid in preference to the boron of the enolate (Scheme 4.10).

One situation in which Lewis acid catalysis is invariably required is in the so-called Mukaiyama reaction in which a 'silyl' enolate (silyl enol ether) is employed as the nucleophilic component (Scheme 4.11). The *anti* selectivity in this case is again thought to be due to the fact that an open transition state is favoured. Other examples are known in which an open transition state leads predominantly to the *syn* product.

4.5 ALTERNATIVE STRATEGY

An alternative and shorter approach to achieving an asymmetric aldol reaction involves attaching a chiral group to boron. This approach has been pioneered by several groups including those of Paterson, Masamune, and Corey. For example, Paterson *et al.* have prepared boron enolates containing either the (+)- or (−)-isopinocampheyl group (eqns 4.15 and 4.16), while Masamune *et al.* have utilized a C_2 symmetric dialkyl boron reagent (Scheme 4.12).

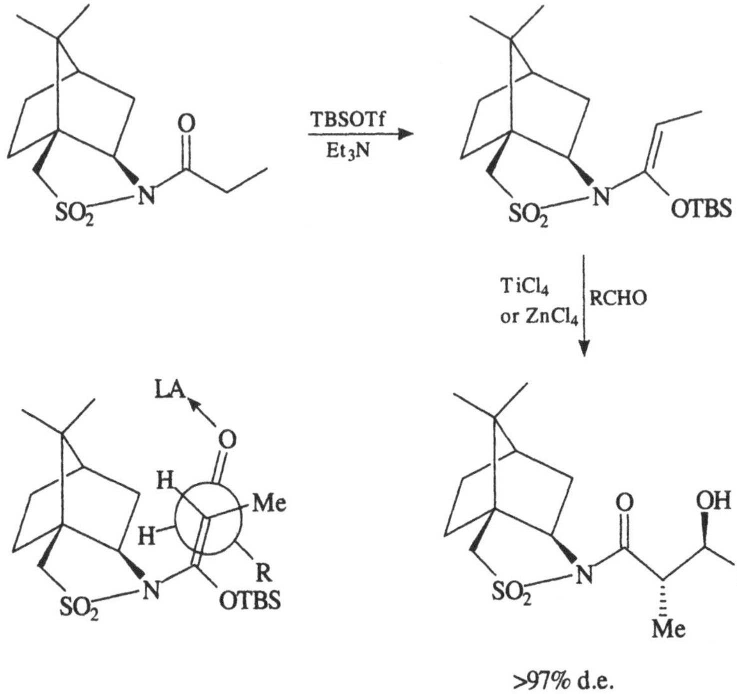

>97% d.e.

Scheme 4.11

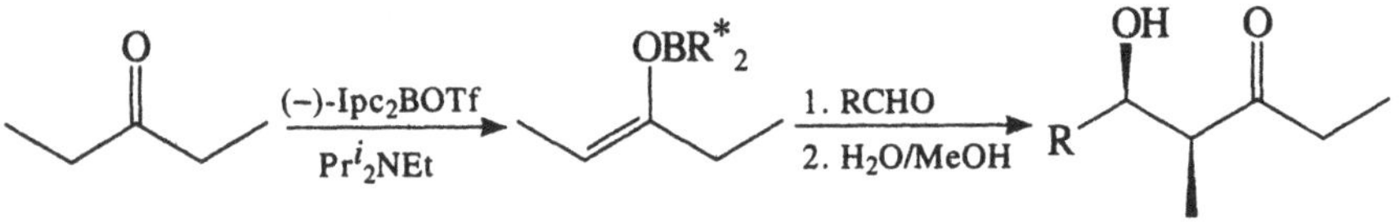

(4.15)

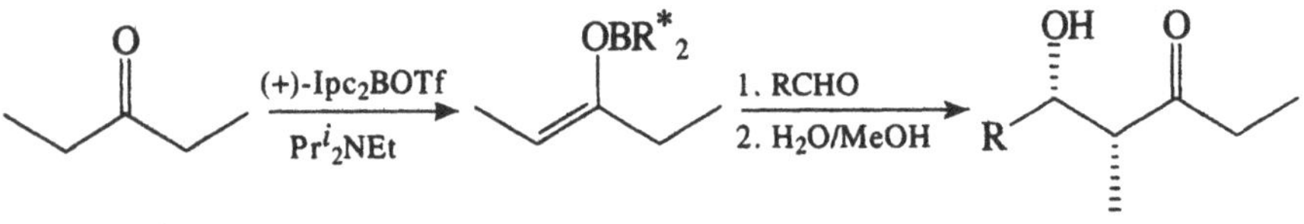

(4.16)

Corey *et al.* have adopted a similar approach by using a chiral boron derivative of 1,2-diamino-1,2-diphenylethane (Scheme 4.13). Since both enantiomers
of the diamine are available all four stereoisomers of the β-hydroxy acid can be
prepared.

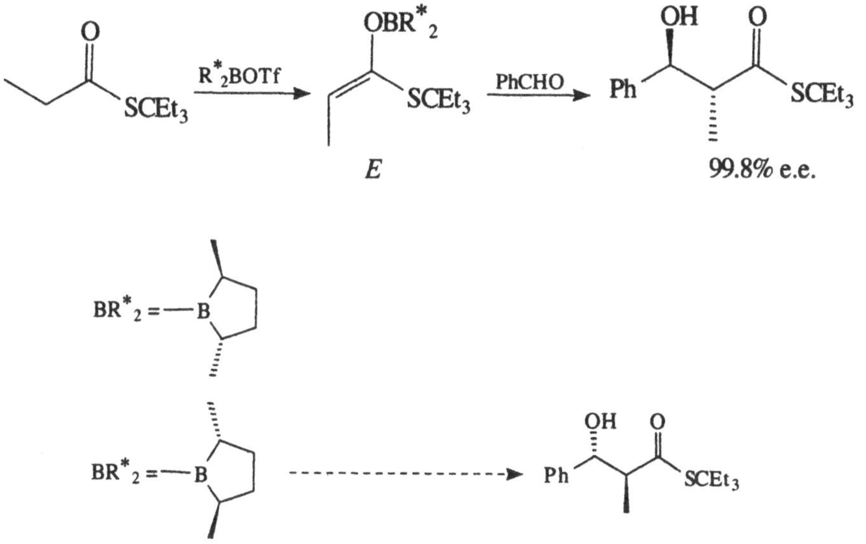

Scheme 4.12

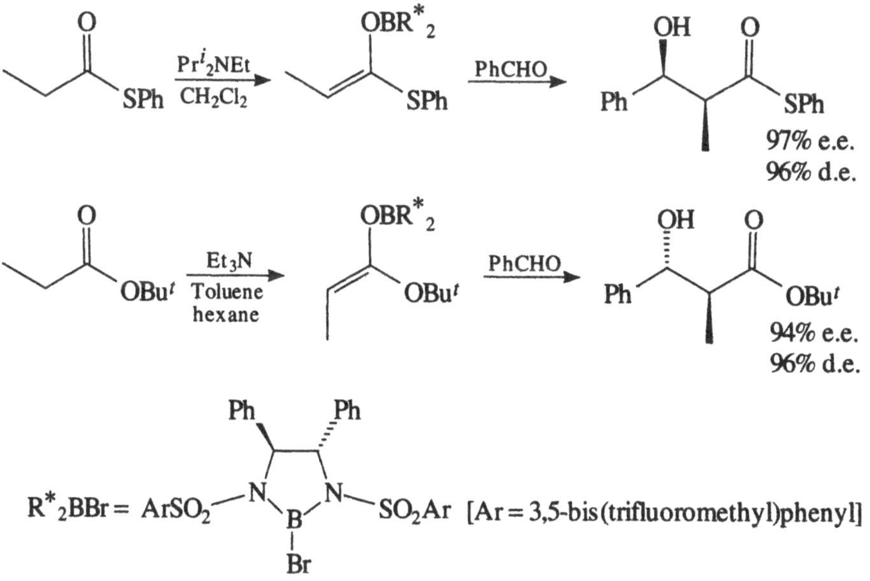

R*₂BBr = ArSO₂—N...B...N—SO₂Ar [Ar = 3,5-bis(trifluoromethyl)phenyl]

Scheme 4.13

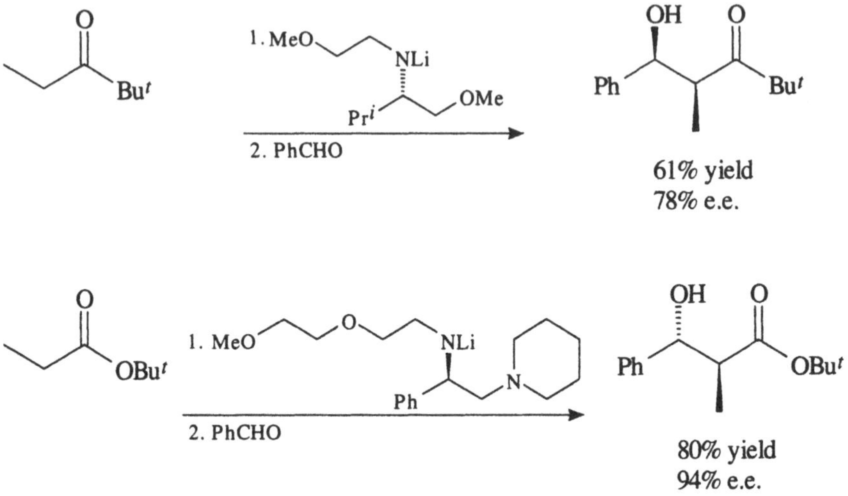

Scheme 4.14

Finally it should be noted that in some instances asymmetric enolate reactions, including the aldol reaction, have also been achieved by using chiral lithium bases (Scheme 4.14).

PROBLEMS

1. Give mechanisms for each of the following reactions and explain why the reactions take a different course:

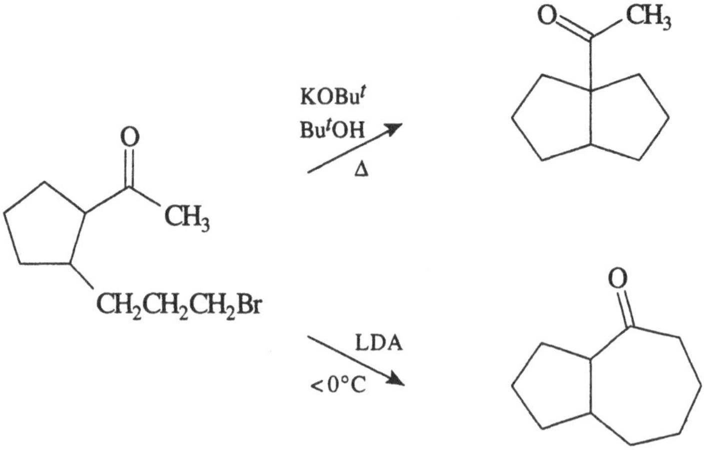

2. Explain why the *anti* aldol product predominates in the following reactions and suggest reasons why the second reaction shows much greater selectivity:

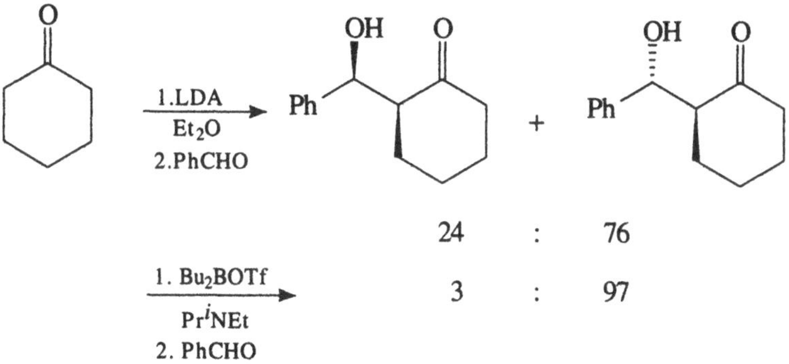

24 : 76

3 : 97

3. A synthesis of (±)-**4** is shown below. Predict the relative configuration of the two stereogenic centres created in **2** and explain your reasoning. Suggest how the scheme could be modified to give an optically active sample of **4**.

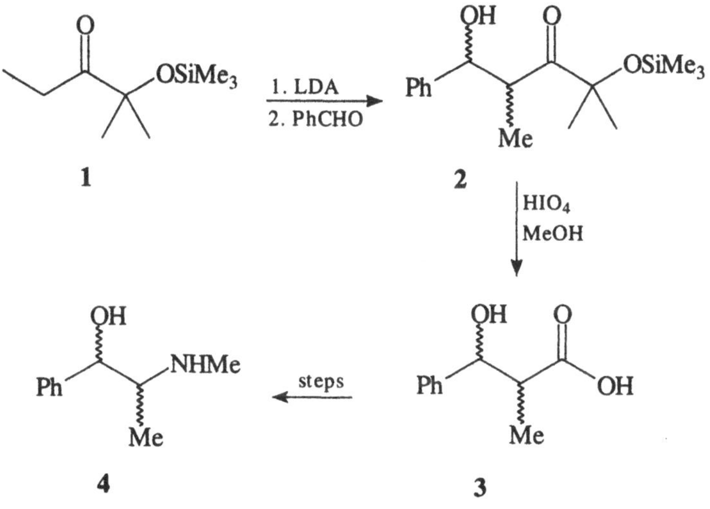

4. How could you prepare the following hydroxy acid from the chiral aldehyde shown using either an aldol or an ene reaction?

5. In the following reaction sequence explain: (i) why the *syn* product **7** is obtained; (ii) the role of the chiral auxiliary X*; and (iii) how the sequence could be modified to produce the enantiomer of **8**.

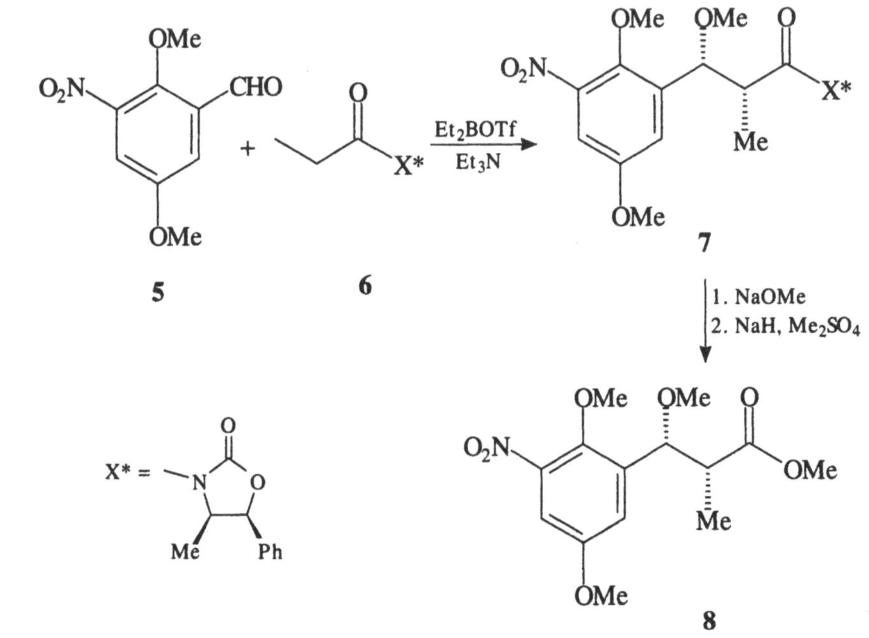

6. The following scheme shows a synthesis of α,α-disubstituted-hydroxy acids involving "chirality transfer". (i) Suggest structures for the intermediates **12** and **13** and hence predict the stereochemistry of the products **14** and **15**. (ii) Explain the role of Bu^{t}CHO in the first step, and explain why MeCHO would be less satisfactory, and HCHO or Me$_2$CO would be totally unsatisfactory for this purpose.

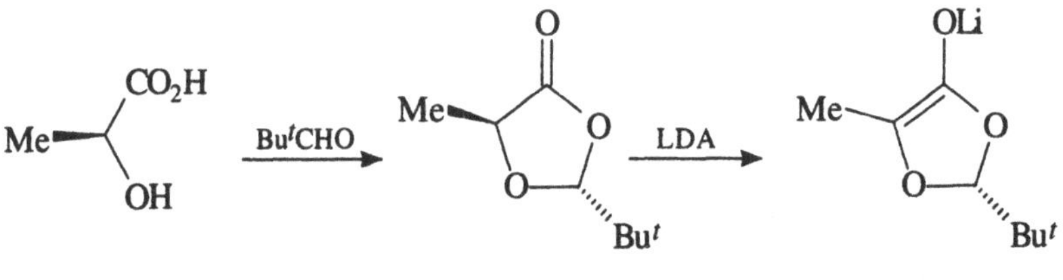

FURTHER READING

G. Procter, *Asymmetric Synthesis*, Oxford University Press, Oxford, 1996, chaps 4 and 5.

Alkylation of enolates

D. A. Evans, M. D. Ennis and D. J. Mathre, *J. Am. Chem. Soc.*, 1982, *104*, 1737.
D. A. Evans and J. Bartoli, *Tetrahedron Lett.*, 1982, *23*, 807.
A. I. Meyers, G. Knaus, K. Kamata and M. E. Ford, *J. Am. Chem. Soc.*, 1976, *98*, 567.
A. I. Meyers, E. S. Snyder and J. J. H. Ackerman, *J. Am. Chem. Soc.*, 1978, *100*, 8186.
D. Enders and H. Eichenauer, *Angew. Chem. Int. Ed.*, 1976, *15*, 549.
A. G. Myers, J. L. Gleason, T. Yoon and D. W. King, *J. Am. Chem. Soc.*, 1997, *119*, 656.

Double asymmetric induction

S. Masamune, W. Choy, J. S. Petersen and L. R. Sita, *Angew. Chem. Int. Ed.*, 1985, *24*, 1.

Anti aldol reactions

D. A. Evans, H. P. Ng, J. S. Clark and D. L. Rieger, *Tetrahedron*, 1992, *48*, 2127.
H. Danda, M. M. Hansen and C. H. Heathcock, *J. Org. Chem.*, 1990, *55*, 173.

Boron enolates

I. Paterson, *Chem. Industry*, 1988, 390.
I. Paterson and M. A. Lister, *Tetrahedron Lett.*, 1988, *29*, 585.
I. Paterson, J. M. Goodman and M. Isaka, *Tetrahedron Lett.*, 1989, *30*, 7121.
R. P. Short and S. Masamune, *Tetrahedron Lett.*, 1987, 2841.

S. Masamune, T. Sato, B.-M. Kim and T. A. Wollmann, *J. Am. Chem. Soc.*, 1986, *108*, 8279.
H. C. Brown *et al.*, *J. Am. Chem. Soc.*, 1989, *111*, 3441.
E. J. Corey and D. H. Lee, *Tetrahedron Lett.*, 1993, 1737.

Use of chiral lithium bases

Y. Landais and P. Ogay, *Tetrahedron Asymmetry*, 1994, *5*, 541.
M. Uragami, K. Tomioka and K. Koga, *Tetrahedron Asymmetry*, 1995, *6*, 701.

ANSWERS

1.

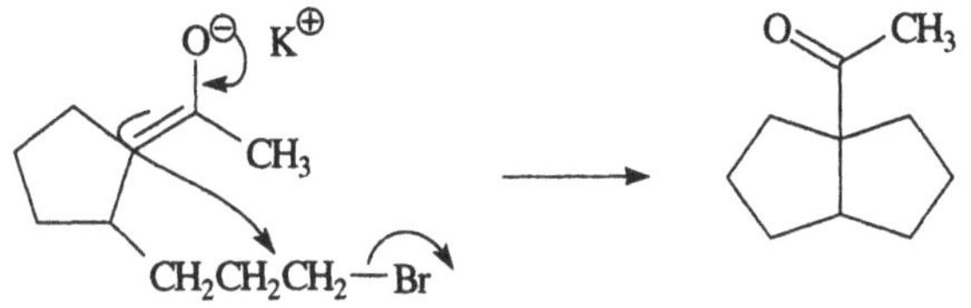

thermodynamic enolate

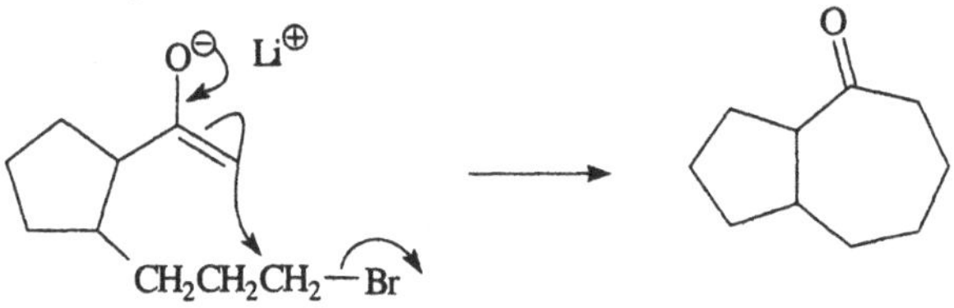

kinetic enolate

2. The *anti* aldol predominates because cyclohexanone is only physically capable of forming an *E* enolate. The first reaction illustrates the poor selectivity provided by the lithium enolate while the second demonstrates the improved selectivity provided by the boron enolate (see M. Majewski and D. M. Gleave, *Tetrahedron Lett.*, 1989, *30*, 5681).

3.

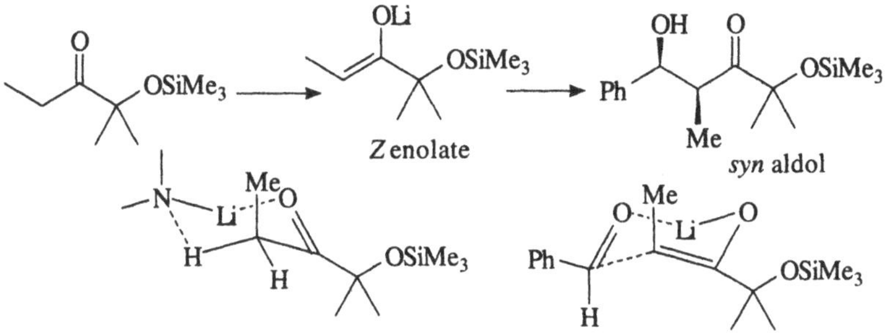

Z enolate   *syn* aldol

To prepare optically active **4** use a chiral enolate, or resolve the acid **3**.

4.

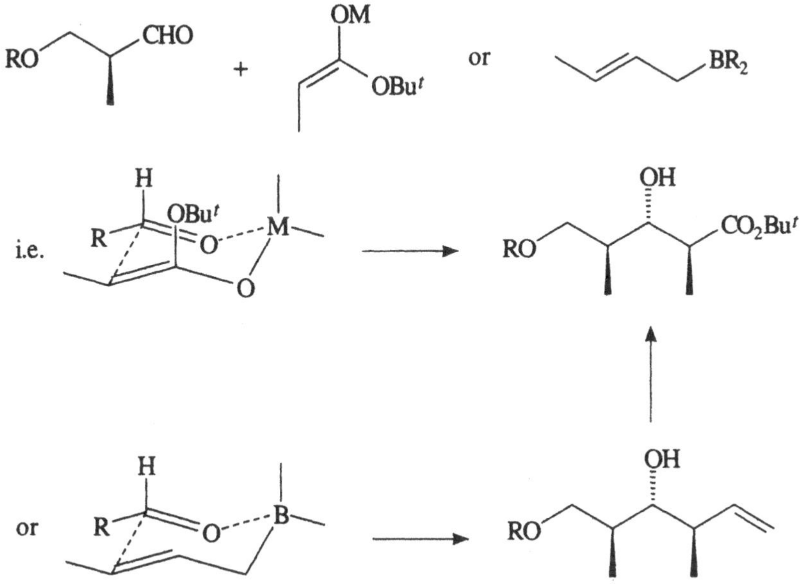

RO—CHO + OM / OBu^t or   BR$_2$

i.e.

$$R \quad H \quad OBu^t \quad O\text{---}M \quad O$$

→ RO— (OH) —CO$_2$Bu^t

↑

or

$$R \quad H \quad O\text{---}B$$

→ RO— (OH) —

5. (i)

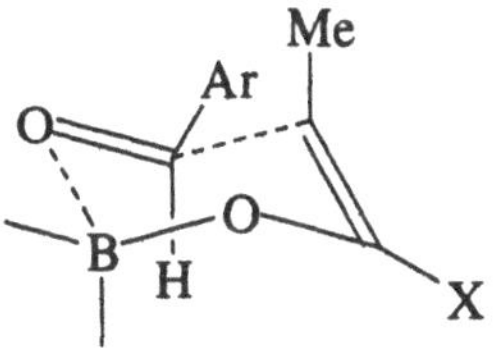

$$Ar \quad Me \quad O \quad B \quad H \quad O \quad X$$

(ii)

$$Ar \quad Me \quad O \quad B \quad H \quad O \quad N \quad Me \quad Ph \quad O \quad O$$

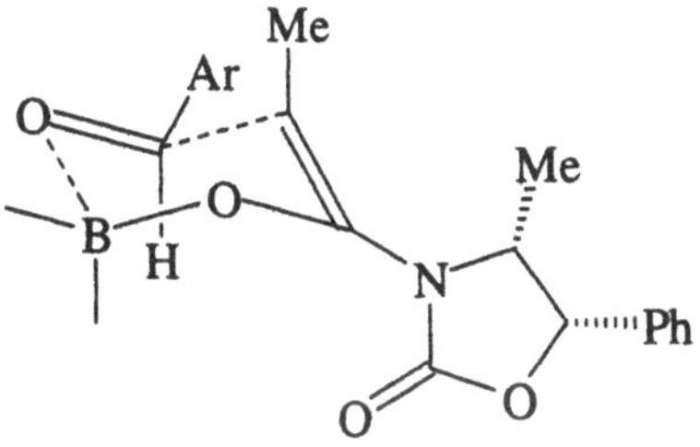

(iii) Use the valine derived chiral auxiliary (see Scheme 4.9).

6. (i)

12	**14**	**13**	**15**

(ii) The chirality induced at the acetal carbon atom in the first step (**9 → 10**) does not form part of the final product but it determines the configuration of the α carbon atom in the final product, the chirality of which is lost in the crucial alkylation/aldol step. Since the methyl group of MeCHO is smaller than the *t*-butyl group it would be less effective in determining the configuration in the enolate reaction. If HCHO or acetone were used the chirality of the original lactic acid would be completely lost when the enolate was formed and hence racemic products would be obtained.

5

Stereoselective Reactions of Alkenes II

In this chapter the hydroboration, hydrogenation, epoxidation, aziridination, cyclopropanation and hydroxylation of alkenes are discussed (Scheme 5.1). Regioselectivity is only an issue in one of these reactions. Furthermore all of them usually display *syn* selectivity.

5.1 STEREOSELECTIVE HYDROBORATION

Hydroboration can be brought about using a variety of reagents including diborane (B_2H_6), borane-THF, borane-dimethyl sulfide (BMS), 9-borabicyclo[3.3.1] nonane (9-BBN), and catecholborane (see Scheme 5.2). The uncatalysed hydroboration of an unsymmetrically substituted alkene is regioselective

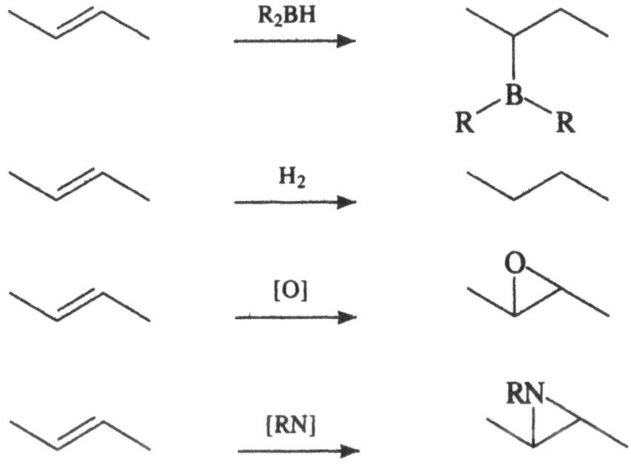

Scheme 5.1

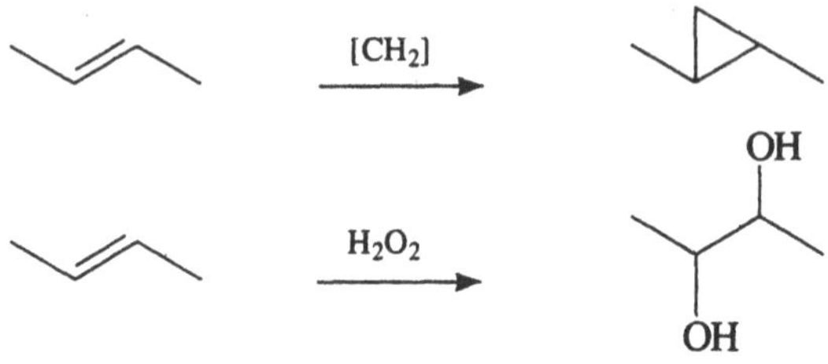

Scheme 5.1 *(continued)*

(non-Markownikov) and stereoselective (*syn*). The regioselectivity appears to be largely determined by steric factors. Furthermore, the hydroboration of a di- or tri-substituted alkene generates up to two new chiral centres and there is therefore ample opportunity to influence the stereoselectivity of the reaction (eqns 5.1–5.3).

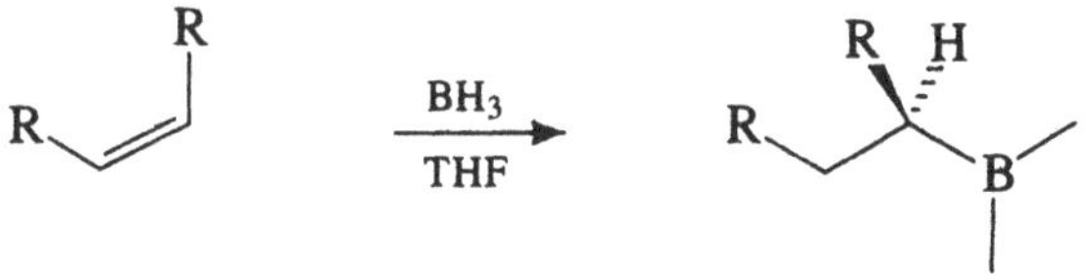

$$(5.1)$$

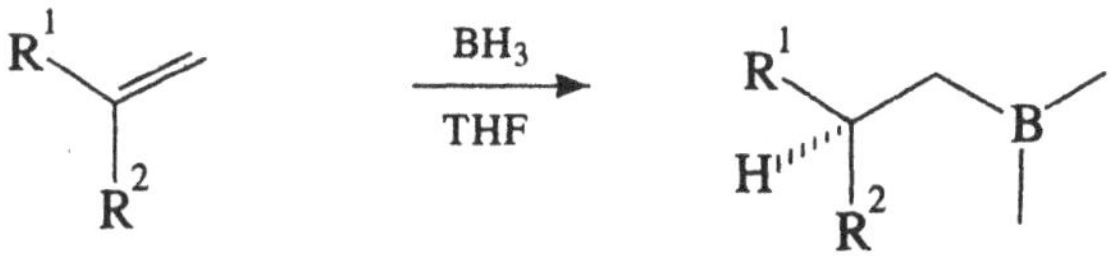

$$(5.2)$$

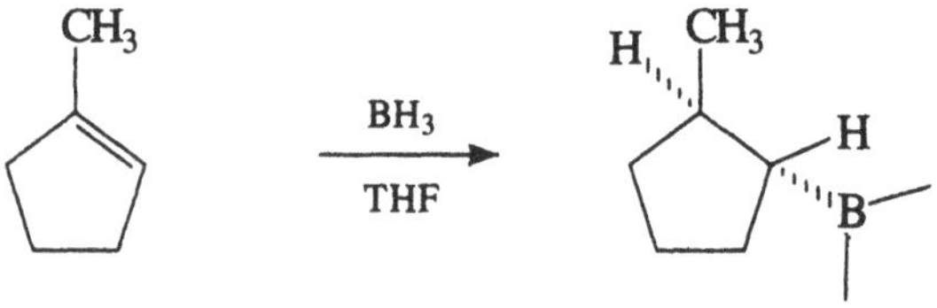

$$(5.3)$$

There are three ways in which the stereoselectivity (and in one case regioselectivity) of this process can be controlled: (i) substrate control; (ii) reagent control; and (iii) catalyst control.

5.1.1 Substrate control

The hydroboration of a chiral substrate proceeds diastereoselectively as illustrated by the following examples (eqns 5.4 and 5.5).

90% d.e.

(5.4)

(5.5)

5.1.2 Reagent control

The two most well known chiral hydroborating agents are mono- and diiso-pinocampheylborane. Diisopinocampheylborane can be prepared directly by hydroboration of α-pinene. Furthermore, both (+)- and (−)-enantiomers of the reagent are available. It is very effective for the asymmetric hydroboration of *cis*-disubstituted alkenes, but less effective in the case of *trans*-disubstituted and trisubstituted alkenes (eqn. 5.6). In contrast, monoisopinocampheylborane which can be prepared by disproportionation of diisopinocampheylborane (eqn. 5.7), gives higher enantiomeric excesses with *trans*-disubstituted alkenes and trisubstituted alkenes (eqns 5.8 and 5.9). Higher enantiomeric excesses can be achieved by recrystallization of the diastereoisomeric di- or trialkylboranes, prior to oxidation.

(+)-α-pinene   (−)-Ipc$_2$BH   (*R*)-(−)-2-butanol

98% e.e.

(5.6)

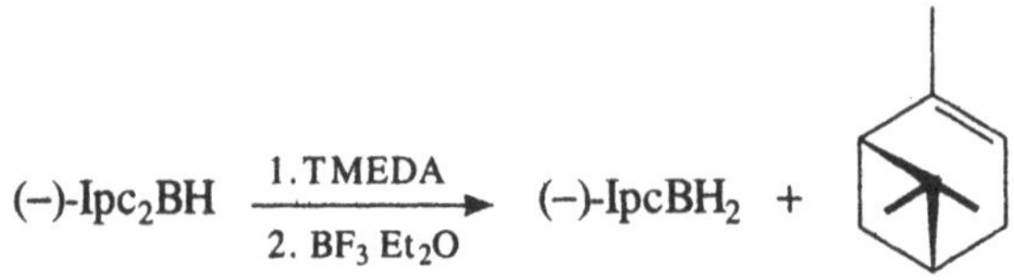

(5.7)

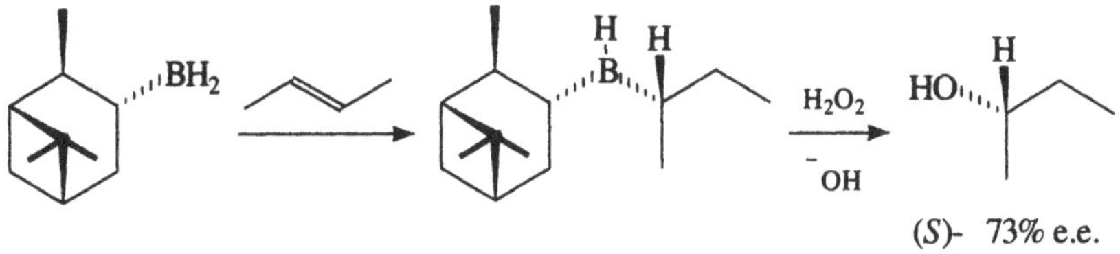

(*S*)- 73% e.e.

(5.8)

66% e.e.

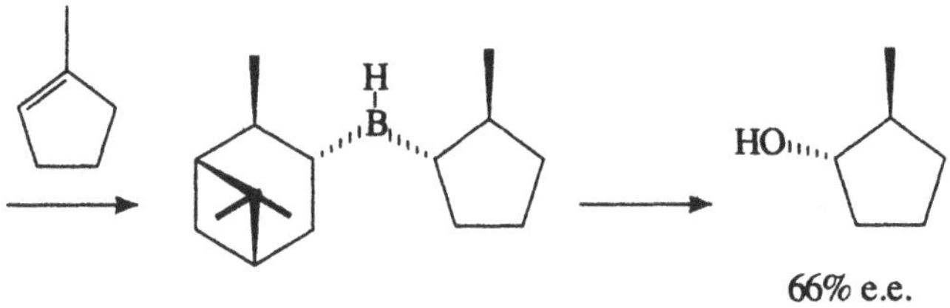

(5.9)

5.1.3 Catalyst control

Hydroboration by catecholborane can be catalysed by rhodium-diphosphine complexes and in some cases even the regioselectivity of the reaction is dramatically changed (Scheme 5.2 and eqn. 5.10). Furthermore, by using chiral

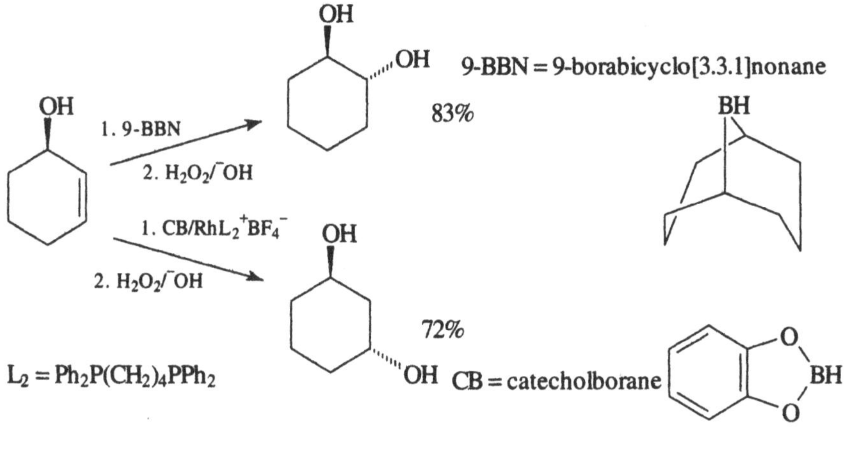

Scheme 5.2

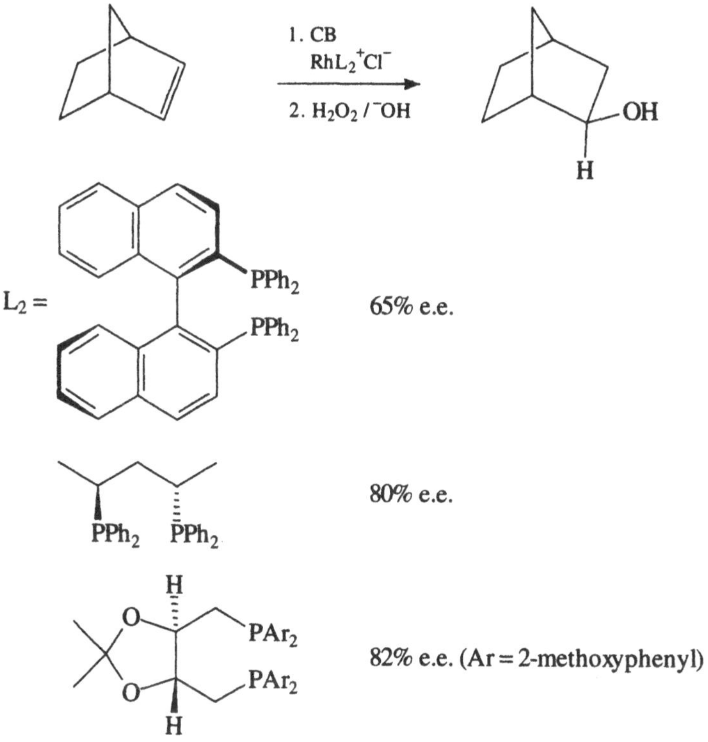

$$L_2 =$$

65% e.e.

80% e.e.

82% e.e. (Ar = 2-methoxyphenyl)

Scheme 5.3

diphosphine ligands, reasonably high enantioselectivities can be achieved (Scheme 5.3).

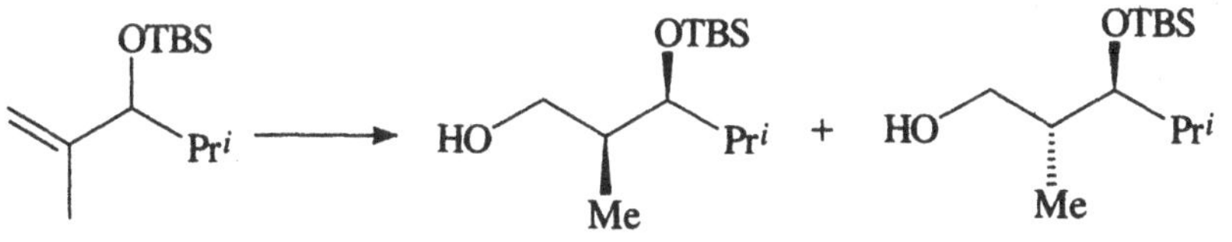

| | 9-BBN | 5 : 95 |
| CB, Rh(PPh$_3$)$_2$Cl | 97 : 3 |

(5.10)

5.2 STEREOSELECTIVE HYDROGENATION

Hydrogenation can be carried out either under heterogeneous conditions using a finely divided metal catalyst, or under homogeneous conditions using a

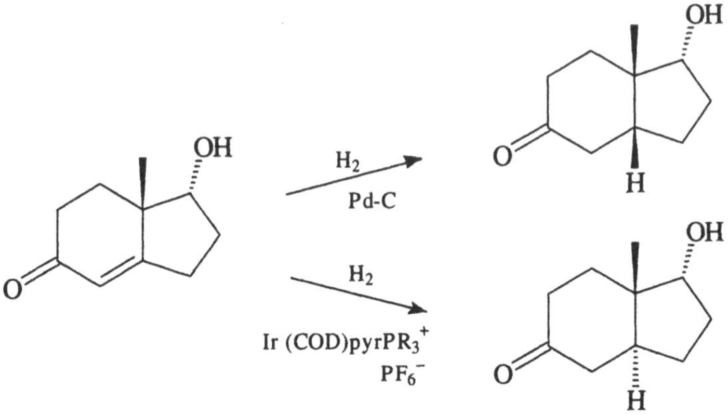

Scheme 5.4

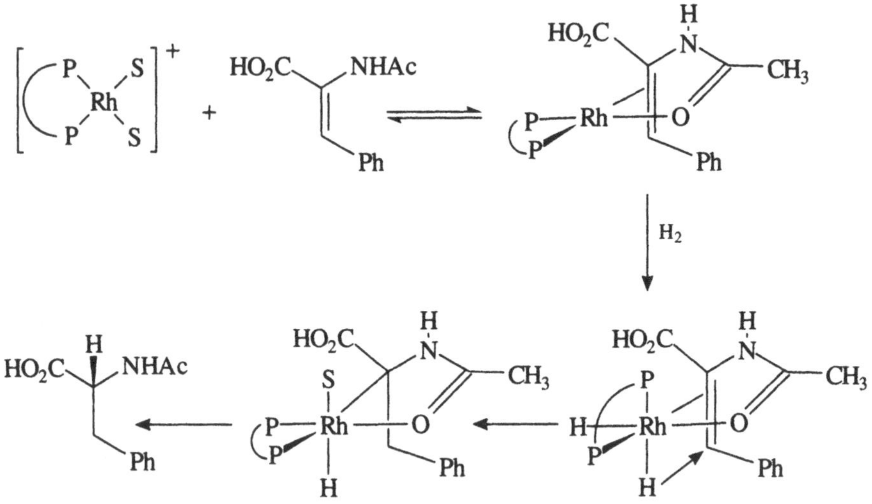

Scheme 5.5

soluble rhodium or iridium catalyst. In these circumstances hydrogenation (like hydroboration) involves *syn* addition. Furthermore absolute diastereoselectivity and enantioselectivity can be achieved either as a result of substrate control (Scheme 5.4), or by using a catalyst containing a chiral phosphine ligand (eqn. 5.11). A possible mechanism for the latter reaction is outlined in Scheme 5.5.

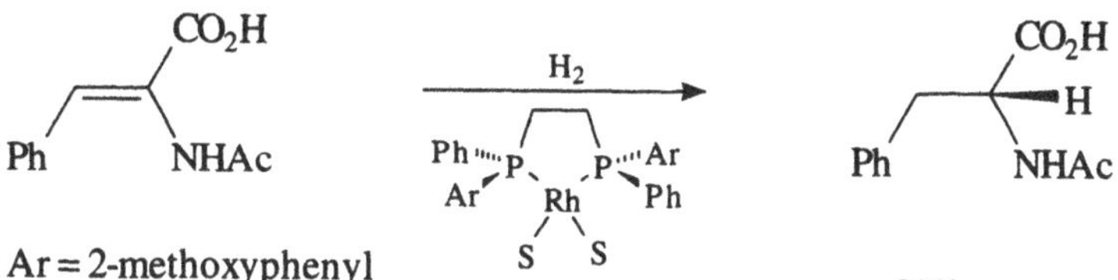

$$\text{Ar} = \text{2-methoxyphenyl}$$
$$\text{S} = \text{solvent}$$

(5.11)

>95% e.e.

5.3 STEREOSELECTIVE EPOXIDATION

There are several different methods for converting alkenes into epoxides. Different protocols can be employed for epoxidizing different kinds of alkenes (e.g. electron rich or electron poor). For example, allylic alcohols are one group of alkenes which can be chemoselectively epoxidized using *tert*-butyl hydroperoxide (TBHP) (eqn. 5.12).

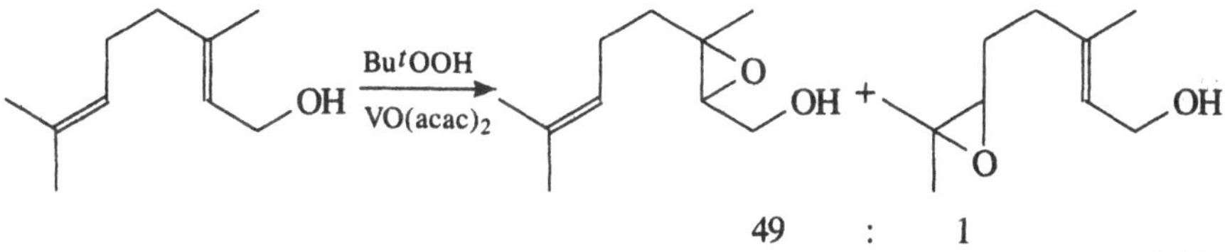

49 : 1

(5.12)

Moreover epoxidation of secondary allylic alcohols using this reagent or a peroxycarboxylic acid is diastereoselective (eqns 5.13 and 5.14, cf. Table 5.1).

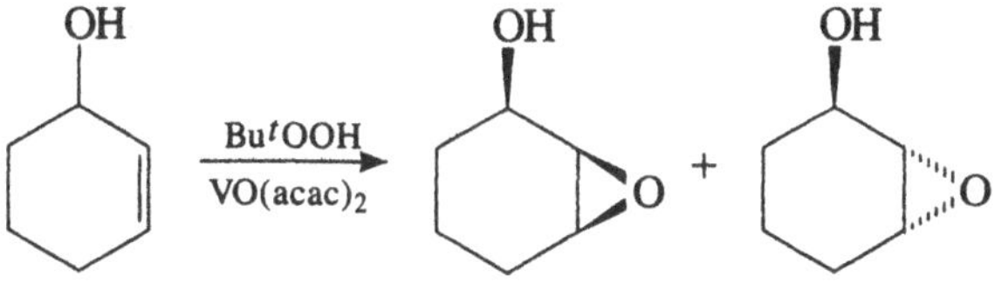

98 : 2
(cf. peracid 92 : 8)

(5.13)

anti   *syn*

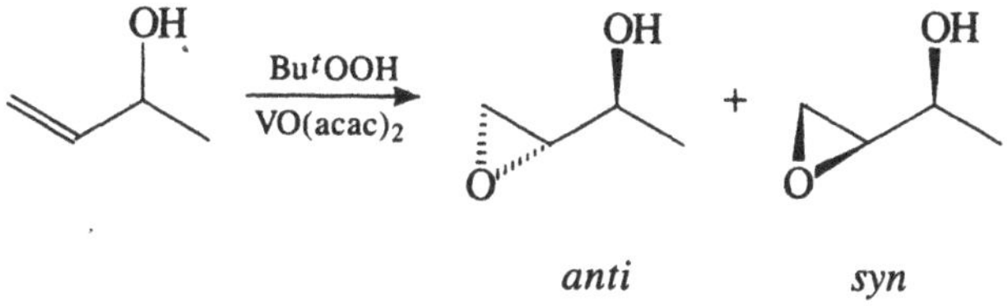

80 : 20
(cf. peracid 40 : 60)

(5.14)

5.3.1 Sharpless epoxidation

The products of the above reactions are of course racemic, unless the substrate used is non-racemic. One of the most useful developments in the field of

Table 5.1 Diastereoselectivity of epoxidation of allylic alcohols using TBHP.

	anti		*syn*	*cf. peracid*
OH structure	80	:	20	(40 : 60)
OH structure	95	:	5	(55 : 45)
OH structure	70	:	30	(37 : 63)
OH structure	30	:	70	(5 : 95)
OH structure	15	:	85	(5 : 95)

Reprinted with permission from *Chemical Reviews*, 1993, *93*, 1307–1370. Copyright 1993 American Chemical Society.

asymmetric synthesis was the discovery of the enantioselective epoxidation of allylic alcohols in the presence of optically active tartrate esters. This reaction is known as "Sharpless epoxidation". The esters most commonly used are (+)- or (−)-diethyl or diisopropyl tartrate (DET and DIPT, respectively). The main attraction of the Sharpless epoxidation procedure is that it affords a single enantiomer of the epoxide in high enantiomeric excess, and in a predictable manner (Scheme 5.6). It can be applied to both primary and secondary allylic alcohols. Furthermore, in the case of racemic secondary allylic alcohols, kinetic resolution occurs (see Chapter 6).

From a purely practical point of view it is found that the addition of 4Å molecular sieves permits the use of only catalytic amounts of the titanium catalyst and the rate of oxidation can be accelerated four to five times by the addition of a drying agent such as calcium hydride and silica gel.

5.3.2 Mechanism of Sharpless epoxidation

The basic mechanism of the reaction is essentially the same whether a vanadium or a titanium catalyst is used (eqns 5.15 and 5.16).

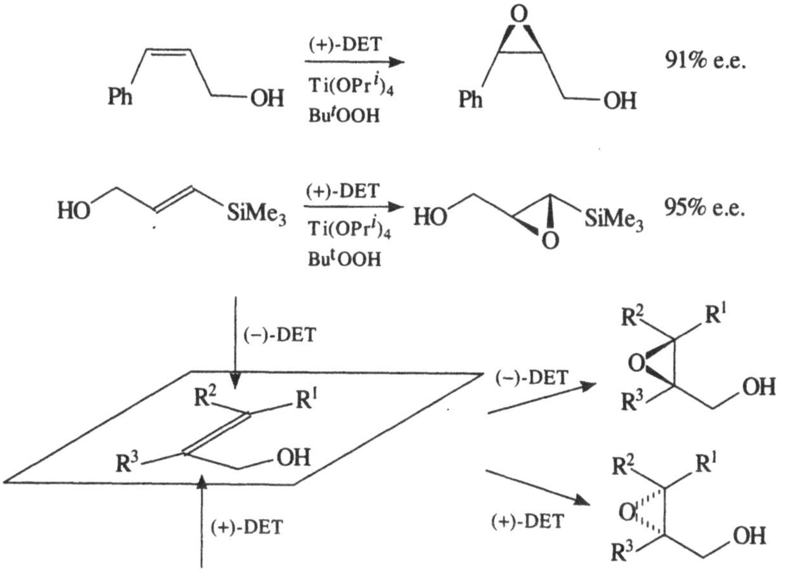

Scheme 5.6

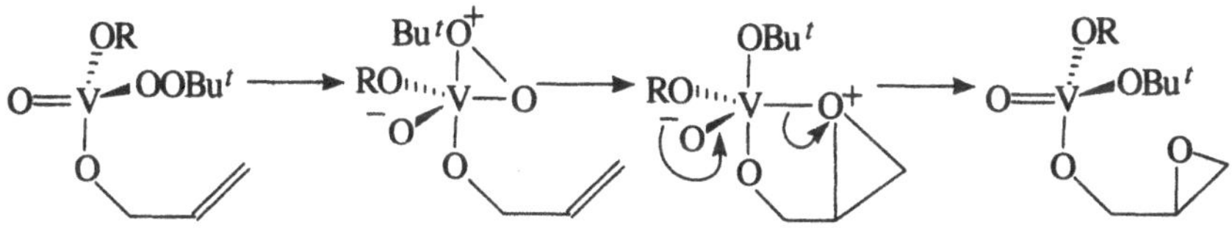

$$(5.15)$$

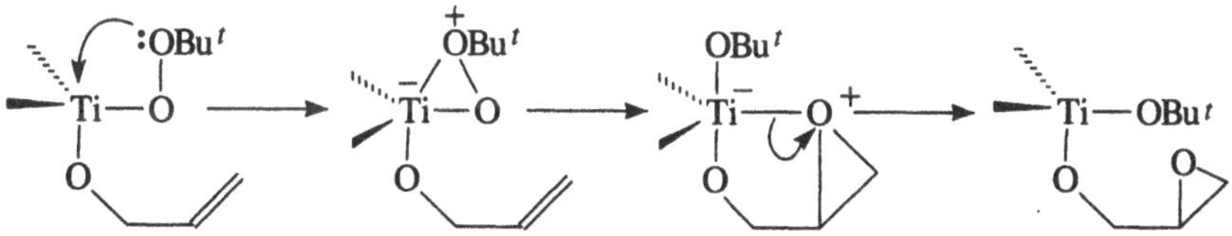

$$(5.16)$$

A number of different models have been proposed to account for the observed stereoselectivity of the reaction. One which shows the allylic alcohol being forced to adopt a conformation in which one face is preferentially exposed to the oxidant is shown in Figure 5.1 (where E is an ester group).

5.3.3 Applications of Sharpless epoxidation

Numerous applications of the Sharpless epoxidation reaction have been reported. The first example illustrates the consequences of *double asymmetric induction*

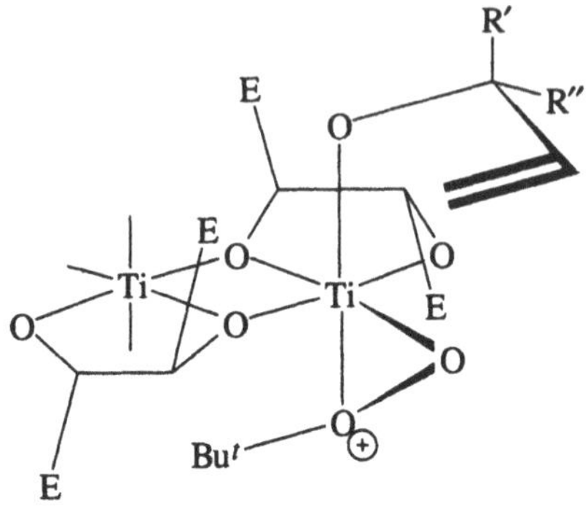

Figure 5.1

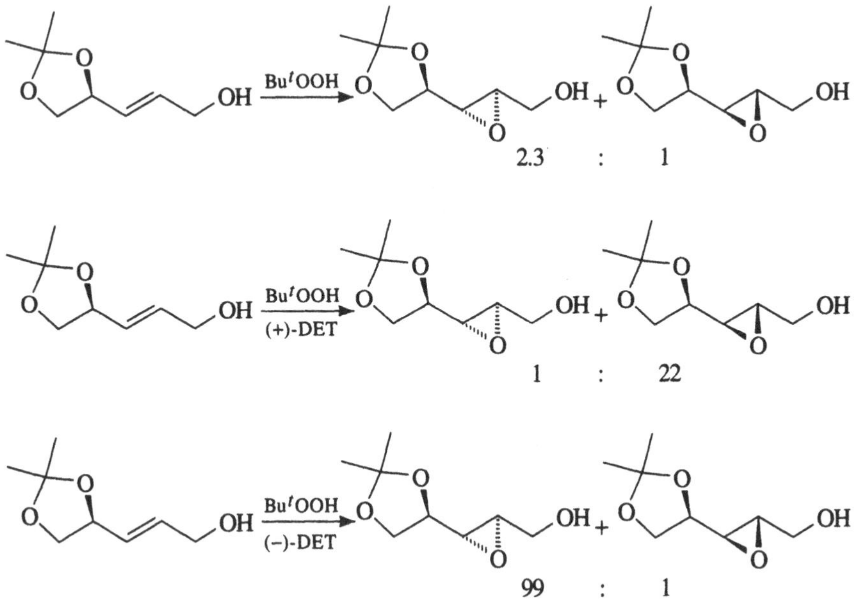

Scheme 5.7

(Scheme 5.7). The combination of a chiral substrate and a chiral reagent lead to the possibility of either "matched" or "mismatched" reactions (cf. Section 4.4).

A second example involves two relatively straightforward syntheses of the β-blocker (S)-propanolol (Scheme 5.8). Both syntheses start from allyl alcohol but use different enantiomers of diethyl tartrate in the epoxidation step.

A third example involves the use of the reaction to synthesize the anti-depressant drug fluoxetine (Scheme 5.9a). The associated schemes (5.9b and

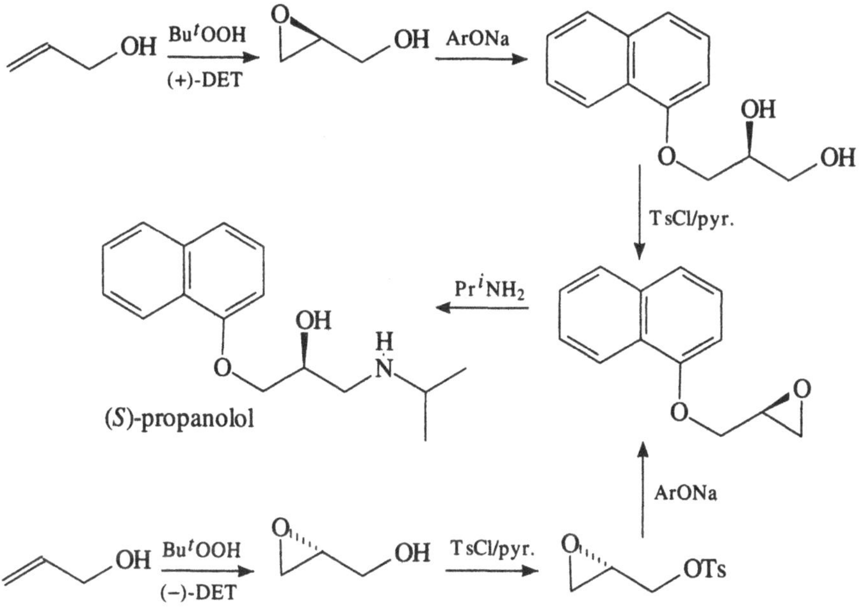

Scheme 5.8

c) illustrate the fact that there are frequently alternative ways of introducing the required chirality into a given target molecule. Thus, two alternative routes to fluoxetine involve asymmetric reduction, one involving a borane reduction of a ketone, and the other an enzyme-catalysed reduction of a β-keto ester.

Finally two routes to a key intermediate in the synthesis of the bryostatins are shown in Scheme 5.10. One route makes use of the Sharpless epoxidation methodology whereas the other uses asymmetric aldol reactions to achieve the same goal.

5.3.4 Jacobsen epoxidation

A second asymmetric epoxidation reaction has been developed by Jacobsen *et al.* This is in principle a much more general procedure since it is not restricted to allylic alcohols. It utilizes a manganese catalyst and uses sodium hypochlorite as the oxidant. It is particularly effective for *cis* alkenes (Scheme 5.11). It is thought that a manganese (IV) oxo species is the actual oxidant.

5.3.5 Epoxidation of $\alpha\beta$-unsaturated carbonyl compounds

Very effective procedures are also available for the epoxidation of $\alpha\beta$-unsaturated carbonyl compounds. The Roberts–Julia method involves the use of a H_2O_2–urea

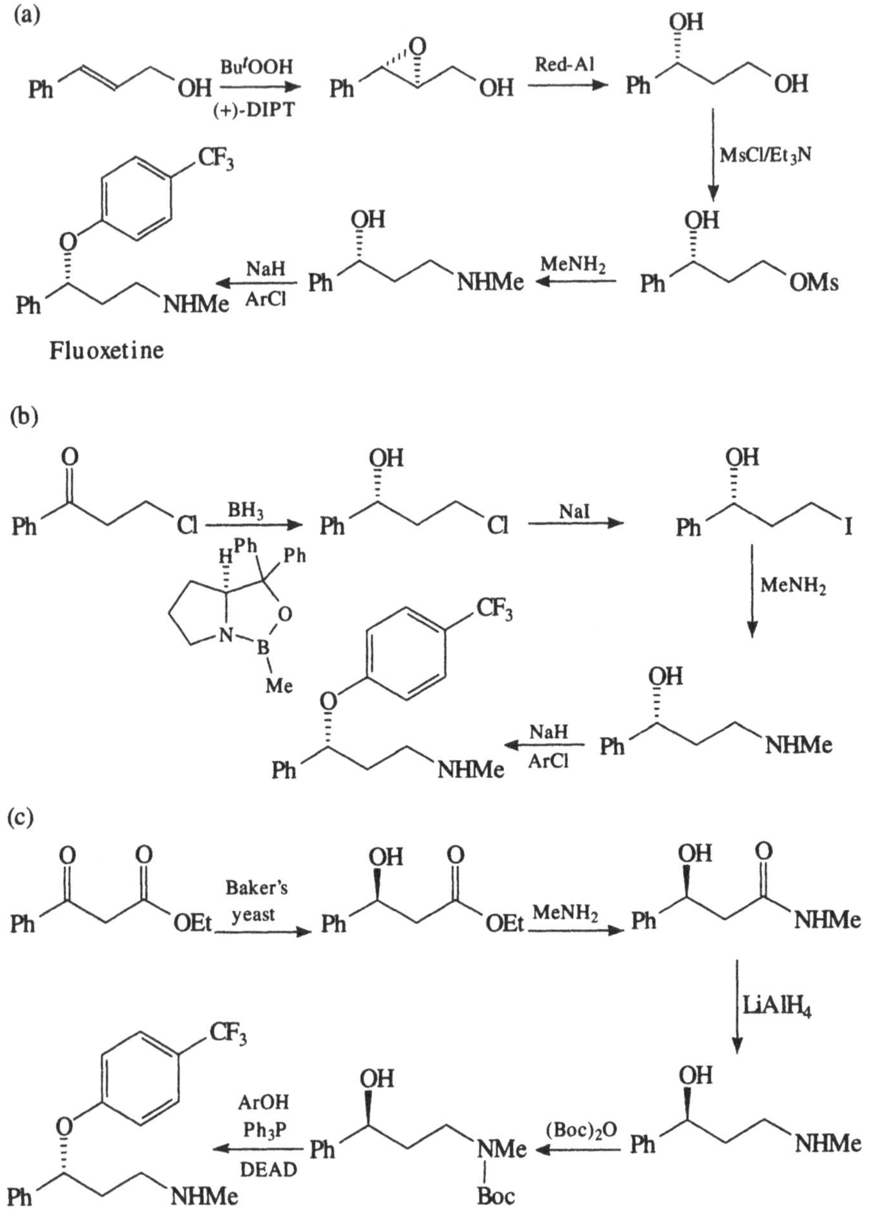
(a)
Ph
OH
Bu'OOH
(+)-DIPT
O
Red-Al
OH
MsCl/Et3N
OMs
MeNH2
NHMe
NaH
ArCl
CF3
Fluoxetine
(b)
Cl
BH3
H
Ph
N
B
Me
NaI
I
MeNH2
NaH
ArCl
(c)
OEt
Baker's
yeast
MeNH2
LiAlH4
(Boc)2O
NMe
Boc
ArOH
Ph3P
DEAD

Scheme 5.9

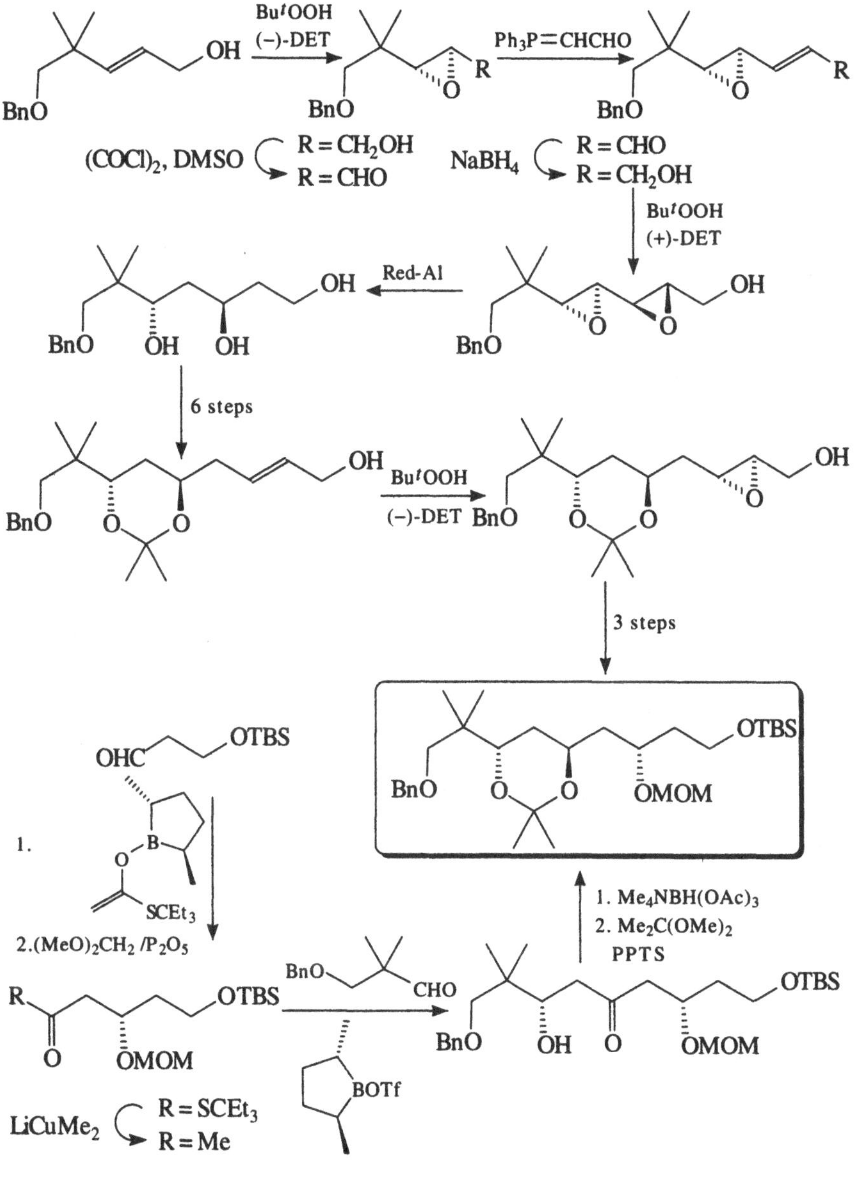

BnO
OH
Bu^tOOH
(–)-DET
R
Ph3P=CHCHO
(COCl)2, DMSO
R = CH2OH
R = CHO
NaBH4
R = CHO
R = CH2OH
Bu^tOOH
(+)-DET
Red-Al
6 steps
Bu^tOOH
(–)-DET
3 steps
OTBS
OMOM
OHC
1.
SCEt3
2.(MeO)2CH2 /P2O5
CHO
BOTf
LiCuMe2
R = SCEt3
R = Me
1. Me4NBH(OAc)3
2. Me2C(OMe)2
PPTS

Scheme 5.10

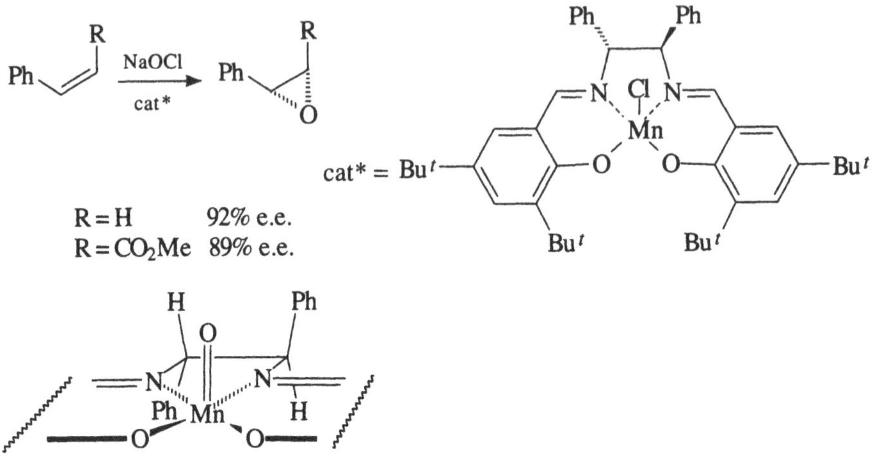

R = H   92% e.e.
R = CO₂Me   89% e.e.

Scheme 5.11

complex in conjunction with a poly-amino acid catalyst (eqn. 5.17), while the use of lithium *tert*-butylperoxide in the presence of (+)-diethyl tartrate provides an alternative approach (eqn. 5.18). Cinchona alkaloid-derived phase-transfer catalysts have also been used with sodium hypochlorite (NaOCl) in a water/toluene two-phase system (Figure 5.2).

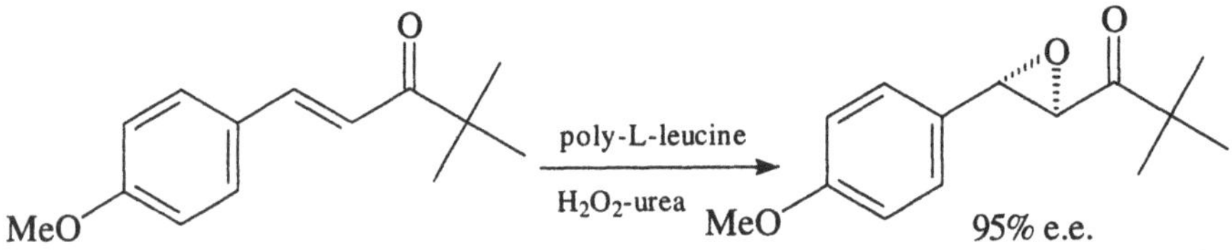

$$(5.17)$$

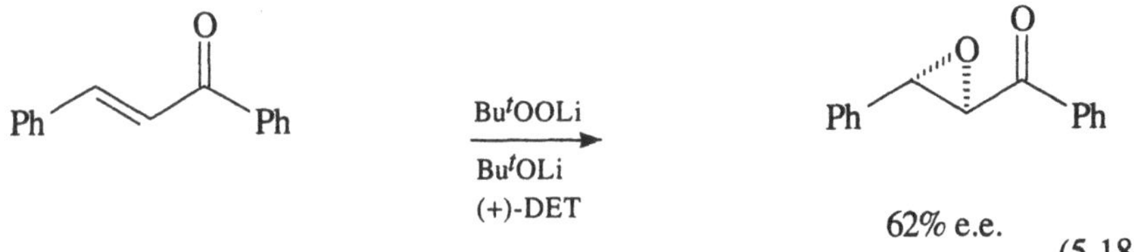

$$(5.18)$$

5.4 CYCLOPROPANATION OF ALLYLIC ALCOHOLS

A reaction with obvious similarities to the Sharpless epoxidation is the Simmons–Smith reaction which leads to diastereoselective cyclopropanation of allylic

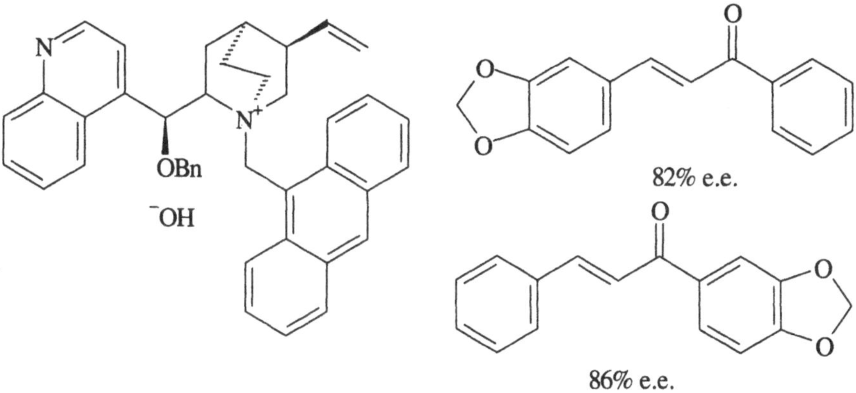

Figure 5.2

alcohols (eqns 5.19 and 5.20).

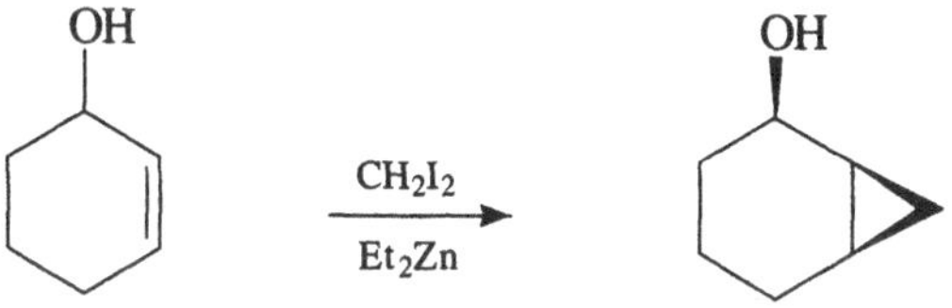

$$\text{(5.19)}$$

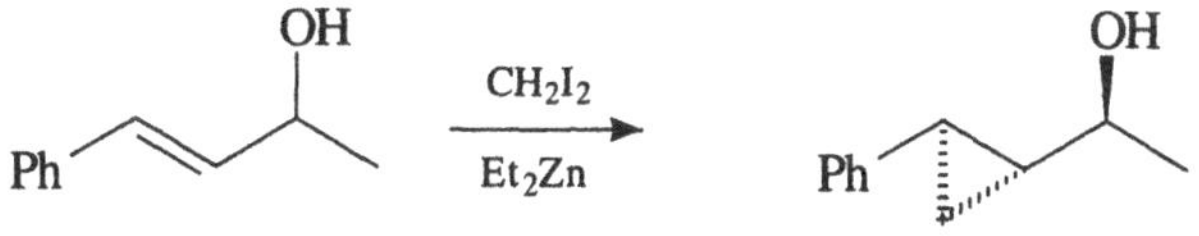

$$\text{(5.20)}$$

In the presence of tartaric acid derivatives enantioselective cyclopropanation occurs (eqn. 5.21).

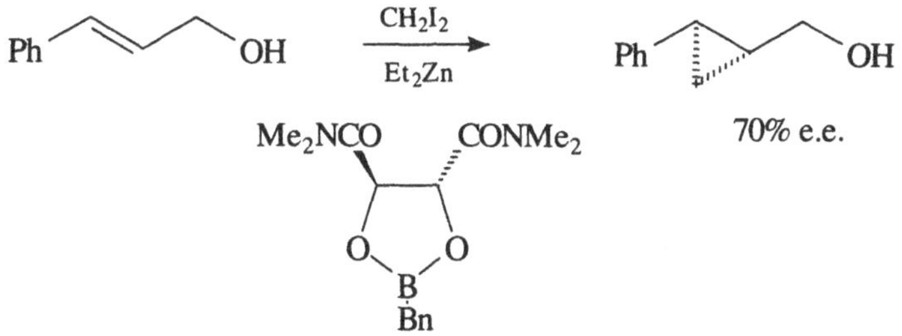

$$\text{(5.21)}$$

5.5 ASYMMETRIC AZIRIDINATION

Not surprisingly work is underway to develop methods for asymmetric aziridination. Just one example is given here in which a chiral 4,4'-disubstituted

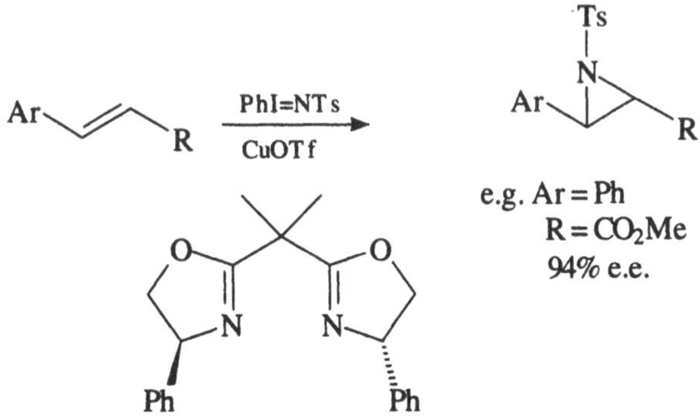

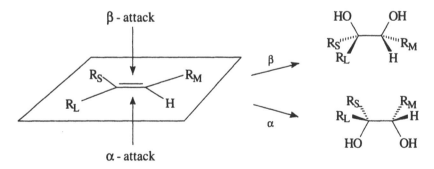

Scheme 5.12

e.g. Ar = Ph
R = CO₂Me
94% e.e.

bis(oxazoline) ligand is attached to a copper catalyst (Scheme 5.12). *N-p-*Toluenesulfonylimino-phenyliodinane is used as the reagent.

5.6 ASYMMETRIC DIHYDROXYLATION

A second very general entry to chiral compounds devised by Sharpless involves asymmetric dihydroxylation of alkenes (Scheme 5.13). In this case two alternative systems are again available affording access to either enantiomer of the diol. The reagent is based on the use of osmium tetroxide and uses either dihydroquinine (DHQ) or dihydroquinidine (DHQD) as the chiral component. In practice the non-volatile osmate $K_2OsO_2(OH)_4$ is used instead of OsO_4.

Once again a well defined model to predict the outcome of the reaction depending upon the steric constraints of the alkene is available (Figure 5.3).

Various derivatives of DHQ and DHQD can be used, but the best ligands have been found to be phthalazine derivatives (Figure 5.4). Only trace amounts of the

Figure 5.3

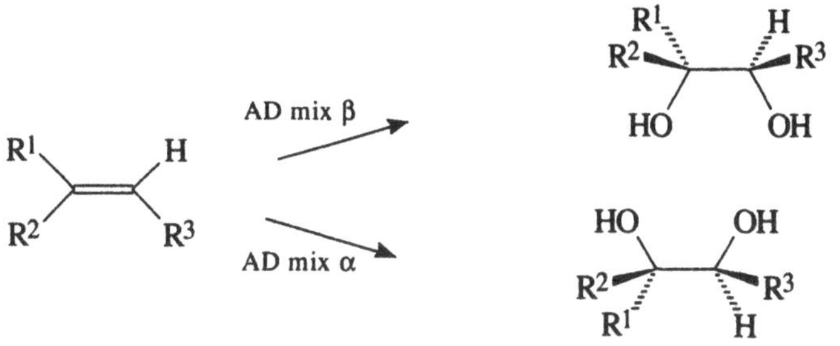

AD mix α = OsO$_4$, K$_3$Fe(CN)$_6$, DHQ (dihydroquinine)
AD mix β = OsO$_4$, K$_3$Fe(CN)$_6$, DHQD (dihydroquinidine)

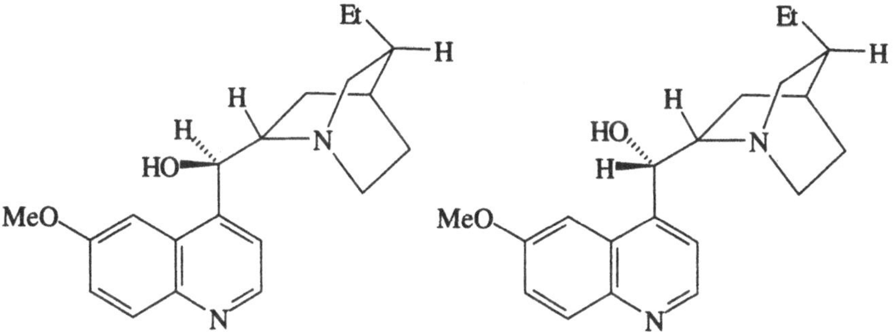

DHQ   DHQD

Scheme 5.13

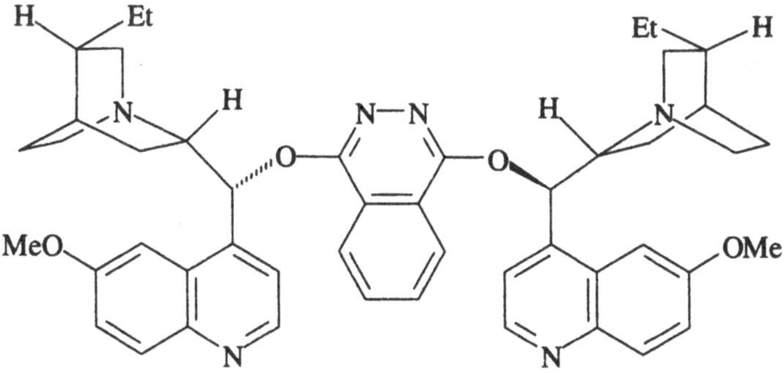

AD mix α

Figure 5.4

ligand and osmium tetroxide are required (0.6% by wt). The main components are potassium ferricyanide and potassium carbonate (99.4% by wt).

5.6.1 Mechanism

The catalytic cycle believed to be involved in the reaction is shown in Scheme 5.14. Potassium osmate provides osmium tetroxide *in situ*, and a complex between this and the chiral ligand brings about enantioselective osmylation of the alkene. Hydrolysis of the resulting osmate ester affords the diol and a reduced osmate species which is then re-oxidised by the stoichiometric oxidant.

5.6.2 Dihydroxylation of aromatic compounds

A related reaction of great interest is the enzyme-catalysed conversion of aromatic compounds to *cis*-diols. These compounds have been used as starting materials for the synthesis of a wide range of target molecules including commercially important compounds (Scheme 5.15).

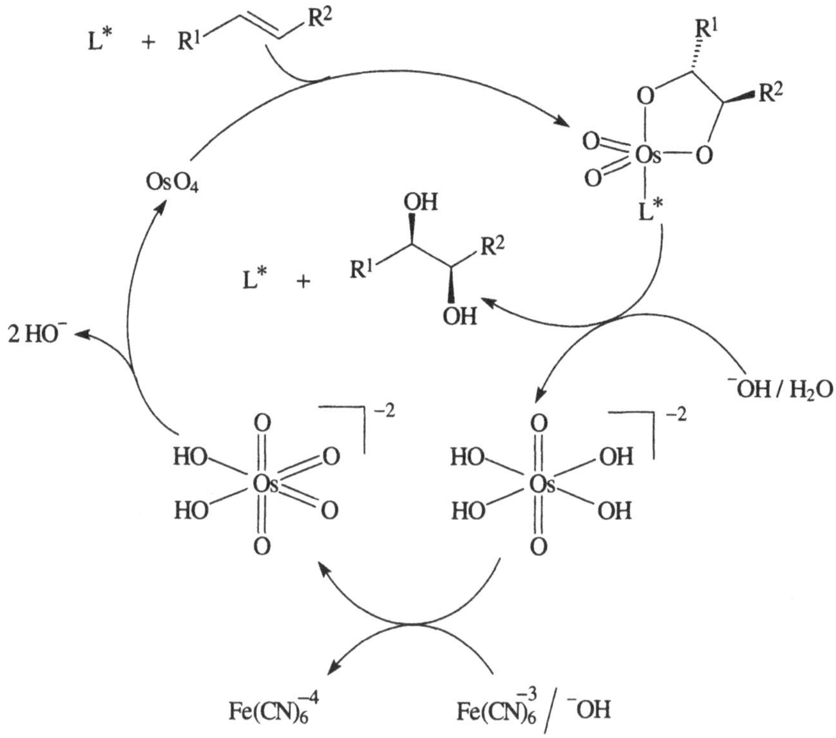

Scheme 5.14

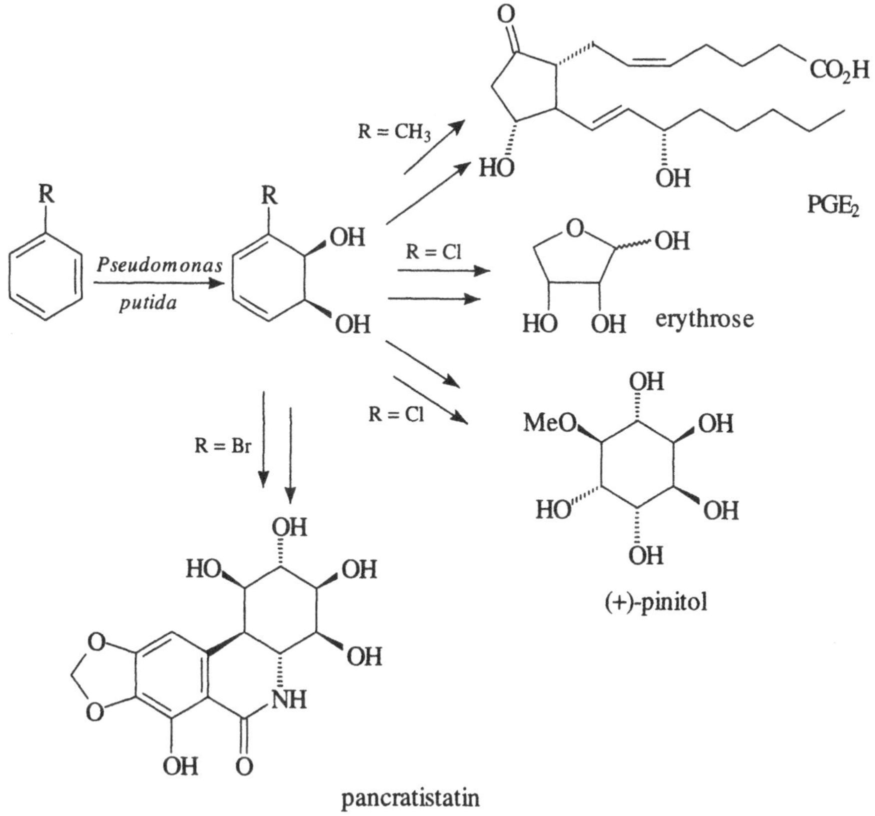

Scheme 5.15

PROBLEMS

1. Suggest a synthesis of **2** from **1** using a Sharpless epoxidation:

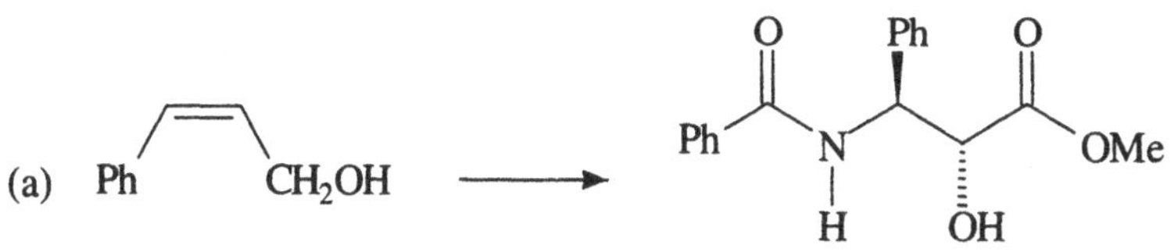

(a)

1

2

2. Supply the missing reagents for steps *c* and *h* in the following sequence and give a mechanism for step *f*.

Reagents: *a*, (EtO)$_2$POCH$_2$CO$_2$Et, NaH; *b*, DIBAL; *d*, H$_3$O$^+$; *e*, TBSCl, Et$_3$N, DMAP; *f*, CH$_3$C(OEt)$_3$, H$^+$; *g*, TBAF.

3. Given that (+)-diop is a chiral diphosphine, while dppb is achiral, use the information provided below to predict the approximate ratio of the two dipeptides obtained when (i) (+)-diop, and (ii) (−)-diop, are used as ligands in the hydrogenation of **A**.

Ph — Ph —
AcNH   CONH   CO₂H
A

[Rh(dppb)]⁺ | H₂

Ph — Ph — + Ph — Ph —
AcNH   CONH   CO₂H   AcNH   CONH   CO₂H
(S,S)-   1.9   :   1   *(R,S)-*

FURTHER READING

A. H. Hoveyda, D. A. Evans and G. C. Fu, *Chem. Rev.*, 1993, *93*, 1307.

Hydroboration

H. C. Brown, J. V. N. V. Prasad, A. K. Gupta and R. K. Bakshi, *J. Org. Chem.*, 1987, *52*, 310.

K. Burgess, W. A. van der Donk and M. J. Ohlmeyer, *Tetrahedron Asymmetry*, 1991, *2*, 613.

Hydrogenation

C. R. Landis and J. Halpern, *J. Am. Chem. Soc.*, 1987, *109*, 1746.

R. Noyori, M. Ohta, Y. Hsiao, M. Kitamura, T. Ohta and H. Takaya, *J. Am. Chem. Soc.*, 1986, *108*, 7117.

H. Takaya, T. Ohta, N. Sayo, H. Kumobayashi, S. Alutagawa, S. Inome, I. Kasahara and R. Noyori, *J. Am. Chem. Soc.*, 1987, *109*, 1596.

J. M. Brown, *Angew. Chem. Int. Ed.*, 1987, *26*, 190.

Sharpless epoxidation

D. Schinzer, in *Organic Synthesis Highlights* (ed. J. Mulzer), VCH Publ., Weinheim, 1991, p. 3.

E. J. Corey, *J. Org. Chem.*, 1990, *55*, 1693.

J. M. Klunder, S.-Y. Ko and K. B. Sharpless, *J. Org. Chem.*, 1986, *51*, 3710.

Syntheses of fluoxetine

Y. Gao and K. B. Sharpless, *J. Org. Chem.*, 1988, *53*, 4081.
E. J. Corey and G. A. Reichard, *Tetrahedron Lett.*, 1989, *30*, 5207.
A. Kumar, D. H. Ner and S. Y. Dike, *Tetrahedron Lett.*, 1991, *32*, 1901.

Synthesis of bryostatins

M. A. Blanchette, M. S. Malamas, M. H. Nantz, J. C. Roberts, P. Somfrai, D. C. Whrite-nour, S. Masamune, M. Kagayama and T. Tamura, *J. Org. Chem.*, 1989, *54*, 2817.

Other epoxidation methods

W. Zhang and E. N. Jacobsen, *J. Org. Chem.*, 1991, *56*, 2296.
L. Deng and E. N. Jacobsen, *J. Org. Chem.*, 1992, *57*, 4320.
M. W. Cappi, W.-P. Chen, R. W. Flood, Y.-W. Liao, S. M. Roberts, J. Skidmore, J. A. Smith and N. M. Williamson, *J. Chem. Soc., Chem. Commun.*, 1998, 1159.
C. L. Elston, R. F. W. Jackson, S. J. F. MacDonald and P. J. Murray, *Angew. Chem. Int. Ed.*, 1997, *36*, 410.
B. Lygo and P. G. Wainwright, *Tetrahedron Lett.*, 1998, *39*, 1599.

Aziridination

D. A. Evans, M. M. Faul, M. T. Bilodeau, B. A. Anderson and D. M. Barnes, *J. Am. Chem. Soc.*, 1993, *115*, 5328.

Dihydroxylation

H. Waldemann, *Organic Synthesis Highlights* (ed. J. Mulzer), VCH Publishers, Weinheim, 1991, p. 9.
H. Becker, M. A. Soler and K. B. Sharpless, *Tetrahedron*, 1995, *51*, 1345.
K. B. Sharpless, W. Amberg, Y. L. Bennani, G. A. Crispino, J. Hartung, K.-S. Jeong, H.-L. Kwong, K. Morikawa, Z.-M. Wang, D. Xu and X.-L. Ahang, *J. Org. Chem.*, 1992, *57*, 2768.

Cyclopropanation

A. B. Charette and J.-F. Marcoux, *Synlett*, 1995, 1205.
A. B. Charette, H. Juteau, H. Lebel and D. Deschenes, *Tetrahedron Lett.*, 1996, *37*, 7925.
A. B. Charette and H. Lebel, *J. Am. Chem. Soc.*, 1996, *118*, 10327.

Dihydroxylation of arenes

T. Hudlicky and A. J. Thorpe, *J. Chem. Soc., Chem. Commun.*, 1996, 1993.

ANSWERS

1.

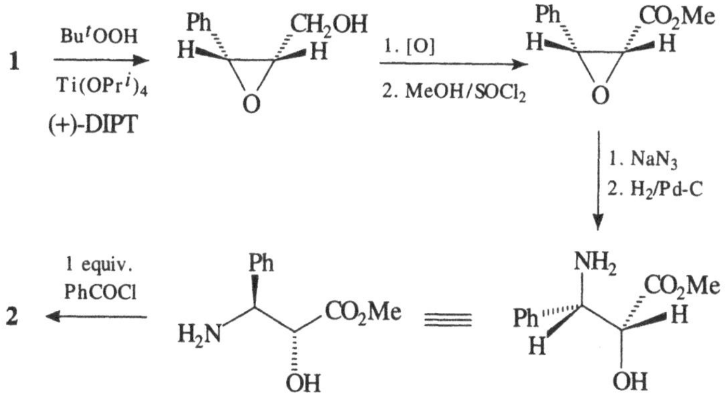

2. *c*, Ti(OPr^i)$_4$, L-(+)-DIPT, Bu^{t}OOH.
   h, Ti(OPr^i)$_4$, D-(−)-DIPT, Bu^{t}OOH.

f

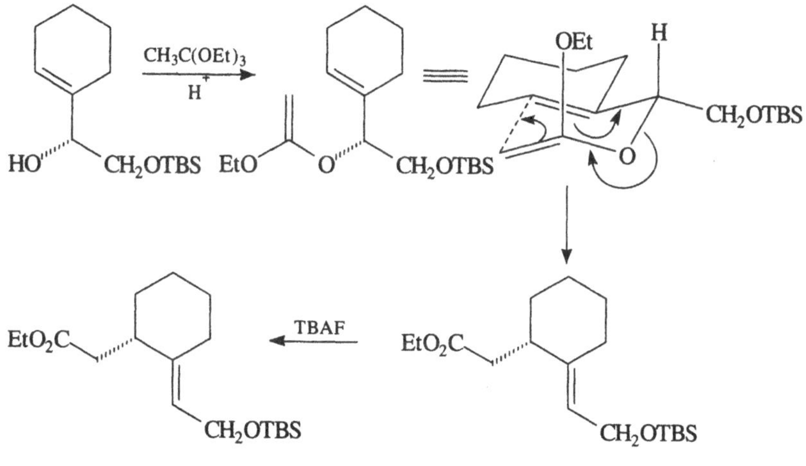

(See J. S. Sabol and R. J. Cregge, *Tetrahedron Lett.*, 1990, *31*, 27.)

3. Predicted values would be (i) 11 × 1.9 = 20.9 : 1 for the "matched" pair (both substrate and catalyst favour formation of the (*S*)-configuration), and (ii) 11 ÷ 1.9 = 5.8 : 1 for the "mismatched" pair ((−)-diop would favour the formation of the (*R*)-configuration by a factor of 11 : 1). The experimental ratios were 16 : 1 and 1 : 4.5, respectively.

6

Kinetic Resolution

Kinetic resolution occurs when one enantiomer in a racemic mixture reacts more rapidly than the other (Figure 6.1). The maximum yield is 100% but the enantiomeric excess of the product decreases as the reaction proceeds. In contrast, the enantiomeric excess of the starting material increases as the reaction proceeds. If the reaction is allowed to go to completion the product is racemic.

Thus kinetic resolution occurs if $k_R \neq k_S$ and the reaction is stopped at some point between 0% and 100% conversion. The ideal situation is one in which only one enantiomer reacts so that at 50% conversion a mixture of 50% of the starting material and 50% of the product can be obtained, both having 100% e.e. Some examples of kinetic resolution are shown below (eqns 6.1–6.4). The dependence of the e.e. of the starting material and the product on the % conversion for the enzyme catalysed hydrolysis of a racemic ester (eqn. 6.1) is illustrated in Figure 6.2.

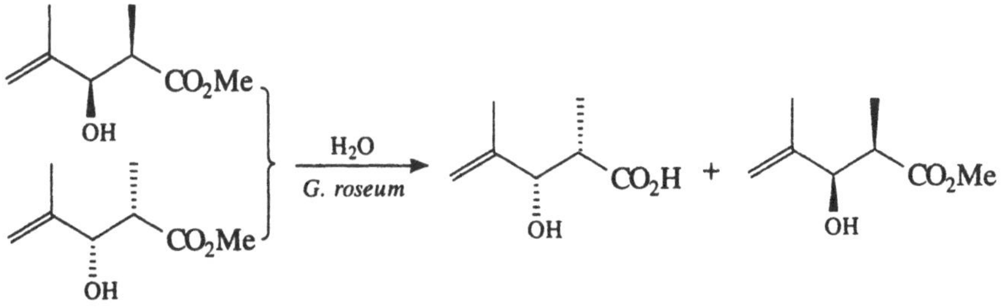

$$(6.1)$$

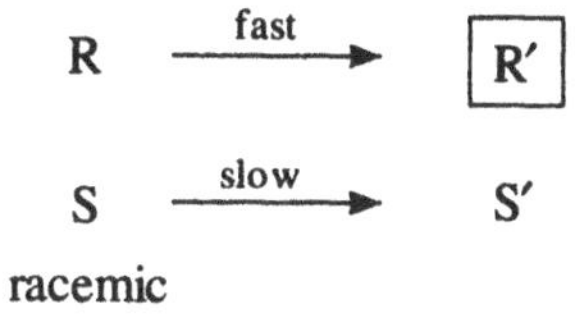

Figure 6.1

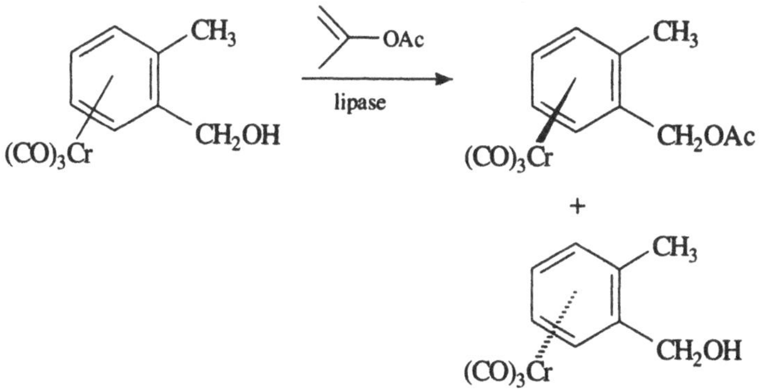

(6.2)

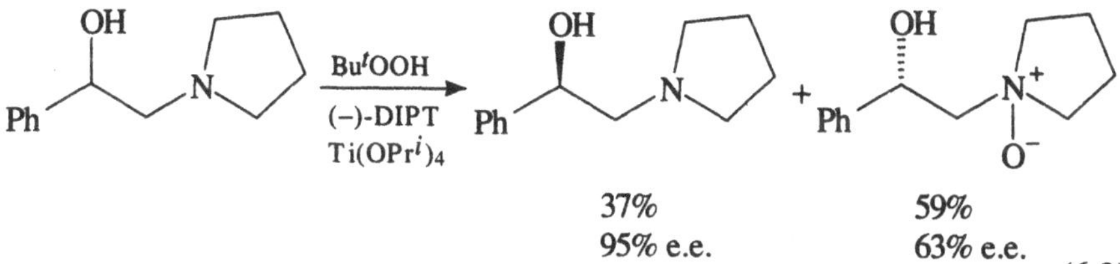

(6.3)

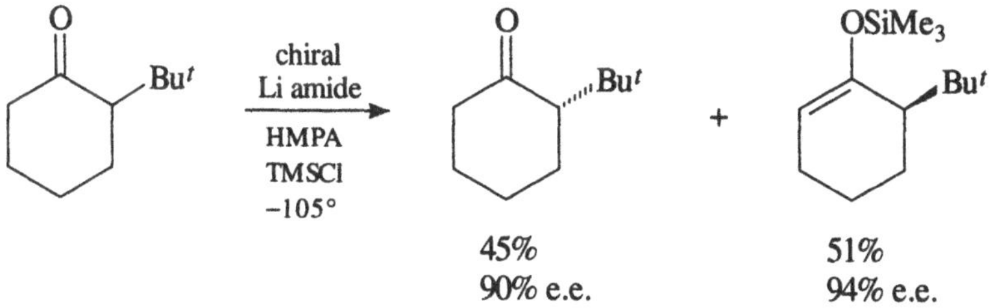

(6.4)

In addition to these and many other examples of straightforward kinetic resolution there are a number of situations in which it occurs in combination with other asymmetric processes, and interesting possibilities then arise.

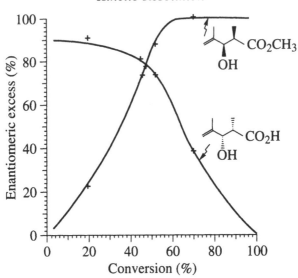

Figure 6.2 Dependence of enantiomeric excess on % conversion for the enzyme catalysed hydrolysis of a racemic ester (eqn. 6.1). (Adapted from C. J. Sih *et al.*, *J. Am. Chem. Soc.*, 1982, *104*, 7294. Reproduced by permission from the American Chemical Society.)

6.1 KINETIC RESOLUTION ACCOMPANIED BY ASYMMETRIC INDUCTION

This situation is very similar to that described in the previous section except that one (or more) new stereogenic centres are created in the kinetic resolution step (Figure 6.3). The maximum yield of the product is again 100% but its enantiomeric excess decreases as the reaction proceeds.

An example is provided by the Sharpless epoxidation of a racemic secondary allylic alcohol (eqn. 6.5). The dependence of the yield and e.e. of the diastereomeric epoxy alcohols on the % conversion is shown in Figure 6.4.

$$
\underset{racemic}{\text{OH}\underset{C_6H_{11}}{\diagup\diagdown}}
\xrightarrow[\substack{\text{0.6 equiv.}}]{\substack{\text{Bu}^t\text{OOH}\\ \text{(+)-DIPT}}}
\underset{\substack{anti\\49\%\\94\%\text{ d.e.}\\>96\%\text{ e.e.}}}{\text{OH}\;C_6H_{11}}
+
\underset{\substack{syn\\\sim 1\%}}{\text{OH}\;C_6H_{11}}
$$

(6.5)

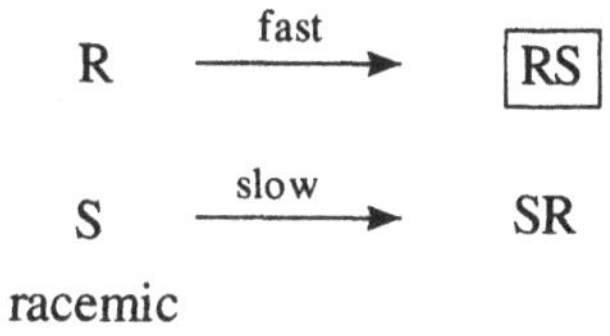

R $\xrightarrow{\text{fast}}$ $\boxed{\text{RS}}$

S $\xrightarrow{\text{slow}}$ SR

racemic

Figure 6.3

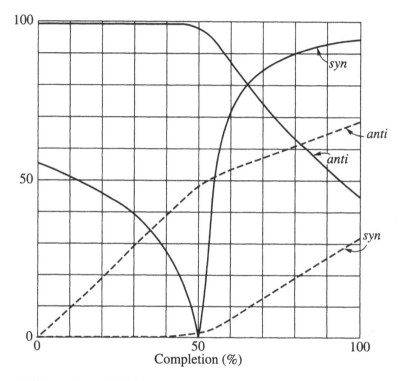

Figure 6.4 Dependence of yield (- - -) and e.e. (———) on % completion for the Sharpless epoxidation of a racemic secondary allylic alcohol (eqn. 6.5). (Adapted from H. B. Kagan and J. C. Fiaud, *Topics in Stereochemistry*, 1988, *18*, 249. Reproduced by permission from John Wiley & Sons, Ltd.)

Since (+)-DIPT requires the R enantiomer to undergo epoxidation from the side shielded by the cyclohexyl group it reacts more slowly than the S enantiomer ($k_S/k_R = 104$) and gives poor diastereoselectivity (Figure 6.5).

A number of other examples of reactions of this type are shown in eqns 6.6–6.11. The major products are shown in each case.

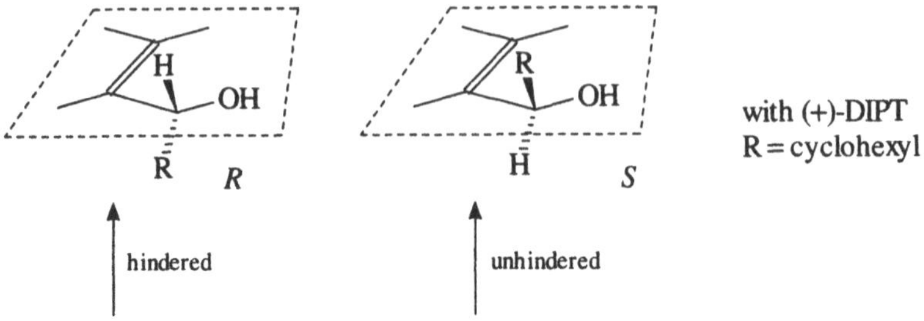

Figure 6.5

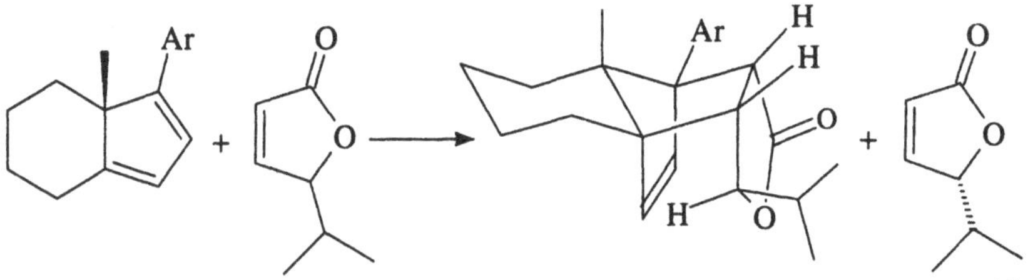

(6.6)

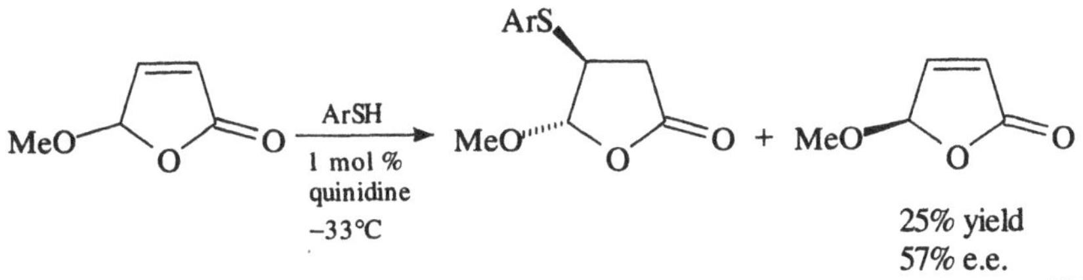

25% yield
57% e.e.

(6.7)

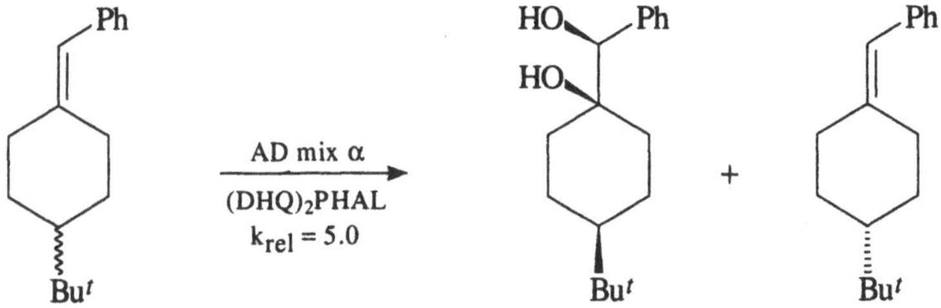

(6.8)

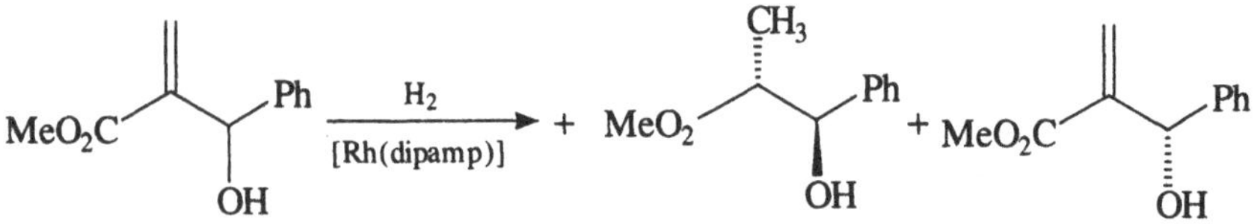

(6.9)

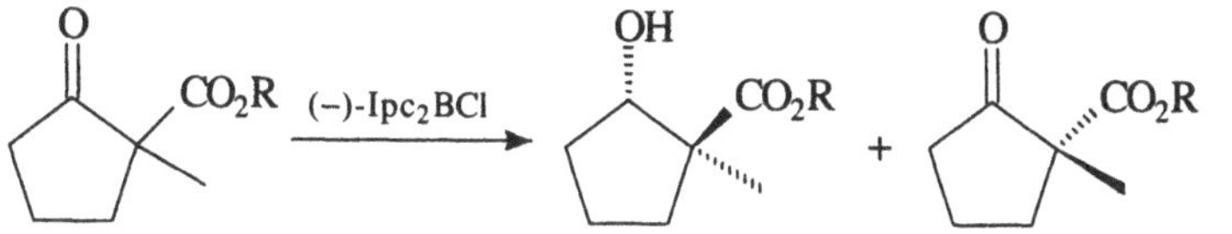

(6.10)

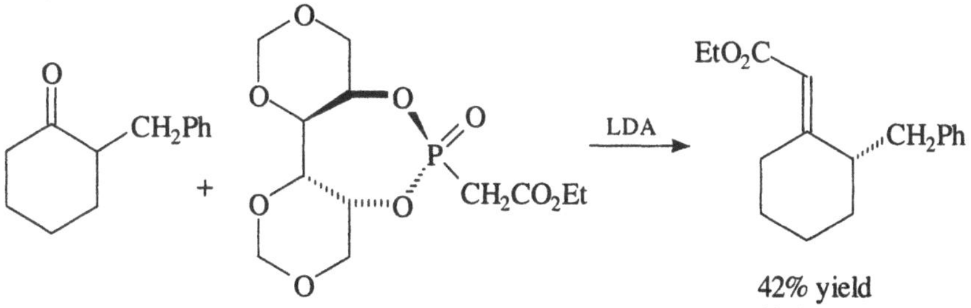

42% yield
89% e.e.

(6.11)

6.2 ASYMMETRIC SYNTHESIS FOLLOWED BY KINETIC RESOLUTION

In this situation an achiral substrate is converted into two enantiomeric products which then undergo kinetic resolution (Figure 6.6). The kinetic resolution enhances the enantiomeric excess of the first formed product. The yield of this product increases to a maximum and then decreases, but its enantiomeric excess continues to increase due to kinetic resolution. A typical example involving the enzyme catalysed hydrolysis of a *meso* diacetate is shown in Scheme 6.1. The % monoacetate and % diacetate composition as a function of the % diol is shown in Figure 6.7.

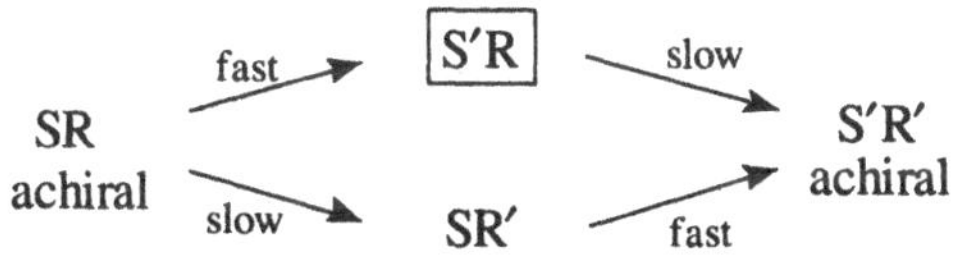

Figure 6.6

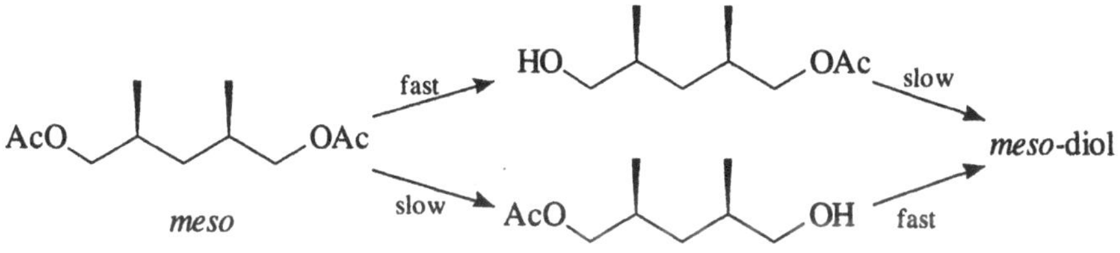

Scheme 6.1

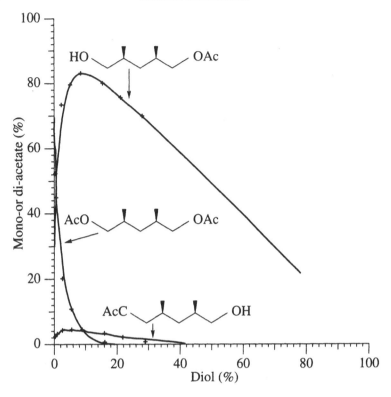

Figure 6.7 Plot of % monoacetate and % diacetate as a function of the % diol for the enzyme catalysed hydrolysis of a *meso* diacetate (Scheme 6.1). (Adapted from C. J. Sih *et al., J. Am. Chem. Soc.*, 1984, *106*, 3695. Reproduced by permission from the American Chemical Society.)

6.3 ENANTIOSELECTIVE HYDROGENATION FOLLOWED BY KINETIC RESOLUTION

A slightly more complicated situation arises in the hydrogenation of the diene (Scheme 6.2) and the reduction of the diketone (Scheme 6.3). In these examples new stereogenic centres are created in two consecutive reactions. It can be seen from Scheme 6.2 that the very high enantioselectivity afforded by the chiral catalyst is the dominant factor in determining the outcome of both steps, with the result that the final product is obtained with 98% d.e. and 95% e.e.

The results shown in Table 6.1 for reduction of the diketone shown in Scheme 6.3 demonstrate that the enantiomeric excess of the α-hydroxyketone increases as the reaction proceeds since the minor enantiomer reacts more rapidly in the kinetic resolution step.

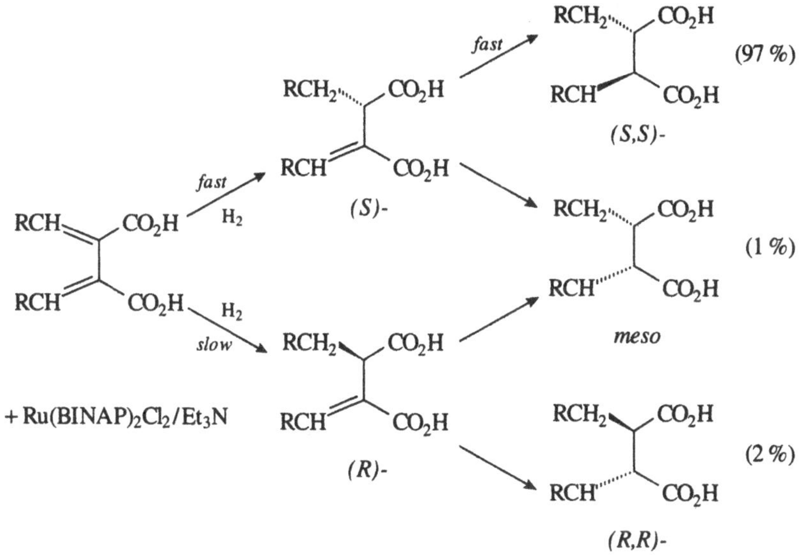

Scheme 6.2

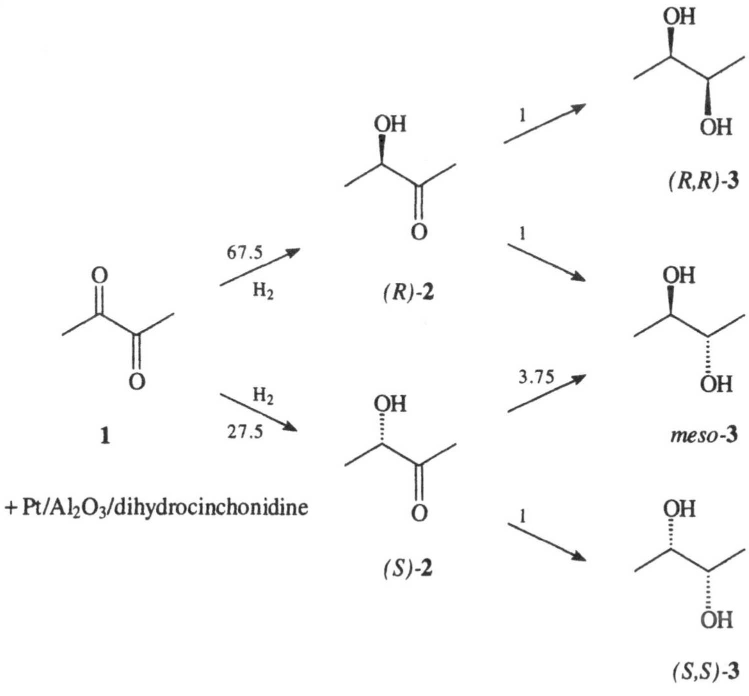

Scheme 6.3

Table 6.1 Analytical data for reduction of the diketone shown in Scheme 6.3.

Time (min)	% (2)	e.e. (2)	% (3)	% meso
15	82	50	9	6
45	45	74	47	31
75	27	85	64	41
90	20	85	71	44
105	13	86	79	49
115	9	90	83	50

Adapted from M. Studer, V. Okafor and H.-U. Blaser, *J. Chem. Soc., Chem. Commun.*, 1998, 1053. Reproduced by permission from the Royal Society of Chemistry.

6.4 SHARPLESS EPOXIDATION OF DIVINYL ALCOHOLS

In this example an achiral (prochiral) bifunctional compound reacts with the creation of one (or more) new stereogenic centres (Scheme 6.4). In the starting material there are four olefinic faces available for epoxidation. Four diastereoisomeric products are formed, but if the rates of the competing reactions are sufficiently different a high degree of selectivity can be achieved.

P_3 is the enantiomer of P_1. If k_1 is greater than k_2, k_3 and k_4 the enantiomeric excess of the product will increase as the reaction proceeds. The yield increases to a maximum and then decreases, but the enantiomeric excess continues to increase

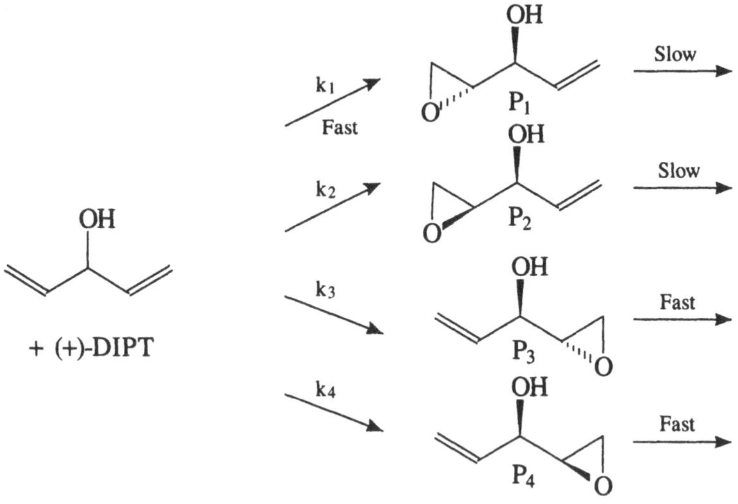

Scheme 6.4

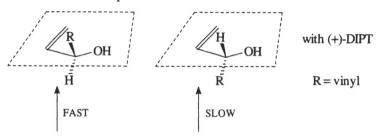

Figure 6.8

Table 6.2 Predicted yield and e.e. data for epoxidation of the divinyl carbinol shown in Scheme 6.4.

S/S_0	Yield P_1/P_0	P_1/P_3	e.e. %
1	0	—	—
0.5	0.48	348	99.4
0.01	0.93	4913	99.96
0.001	0.91	31936	99.994

Adapted from S. L. Schreiber, T. S. Schreiber and D. B. Smith, *J. Am. Chem. Soc.*, 1987, *109*, 1525. Reproduced by permission from the American Chemical Society.

due to kinetic resolution of the product. The selectivity can be understood by considering the reaction on each double bond separately (Figure 6.8).

Calculations based on the relative rates of reaction and stereoselectivities of a typical monoallylic alcohol support the above analysis (Table 6.2).

A practical demonstration of the stereoselectivity which can be achieved is shown in eqn. 6.12.

$$
\begin{array}{ccc}
 & \text{L-(+)-DET} & \\
 & \text{Bu}^t\text{OOH} & 85\% \text{ yield} \\
\xrightarrow{\hspace{2cm}} & & >99\% \text{ e.e.} \\
 & \text{Ti(OPr}^i)_4 & >99\% \text{ d.e.}
\end{array}
\qquad (6.12)
$$

A synthesis of riboflavin making use of this procedure is shown in Scheme 6.5.

Other syntheses of biologically important compounds from divinyl carbinols are shown in outline in Schemes 6.6–6.8.

6.5 SEQUENTIAL KINETIC RESOLUTION

Sequential kinetic resolution can in some cases lead to exceptionally high enantiomeric excesses if the selectivity of both steps reinforce one another

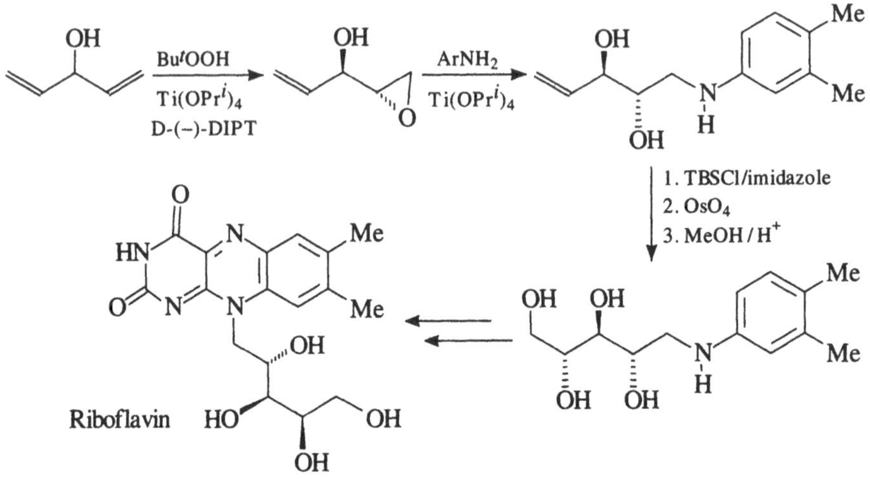

Scheme 6.5

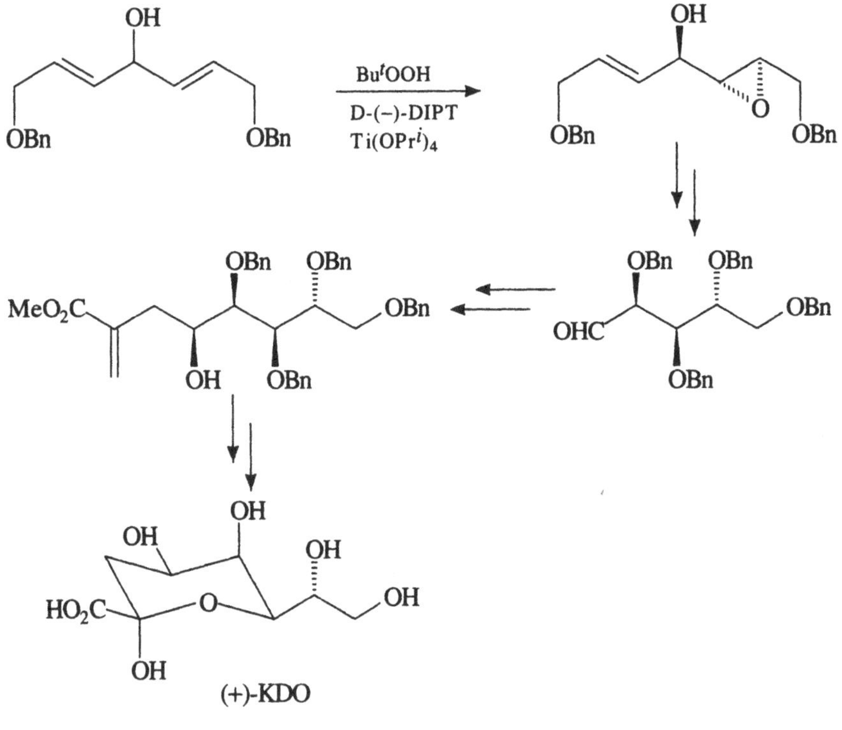

Scheme 6.6

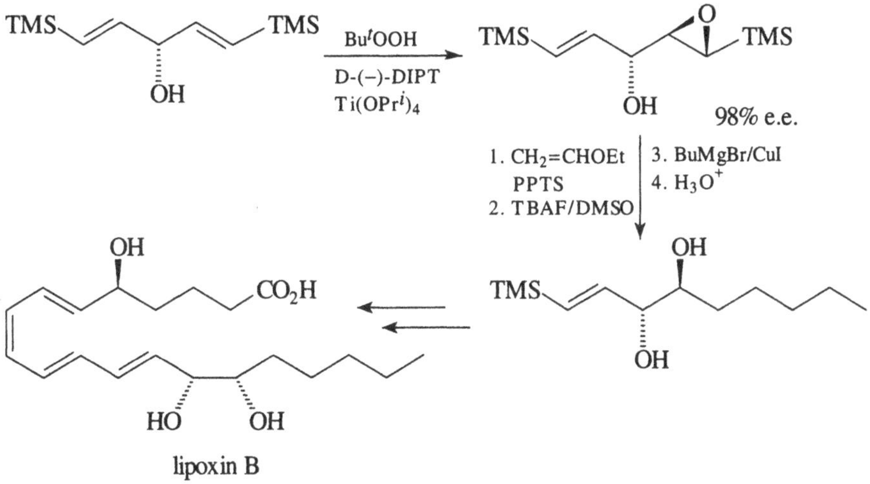

Scheme 6.7

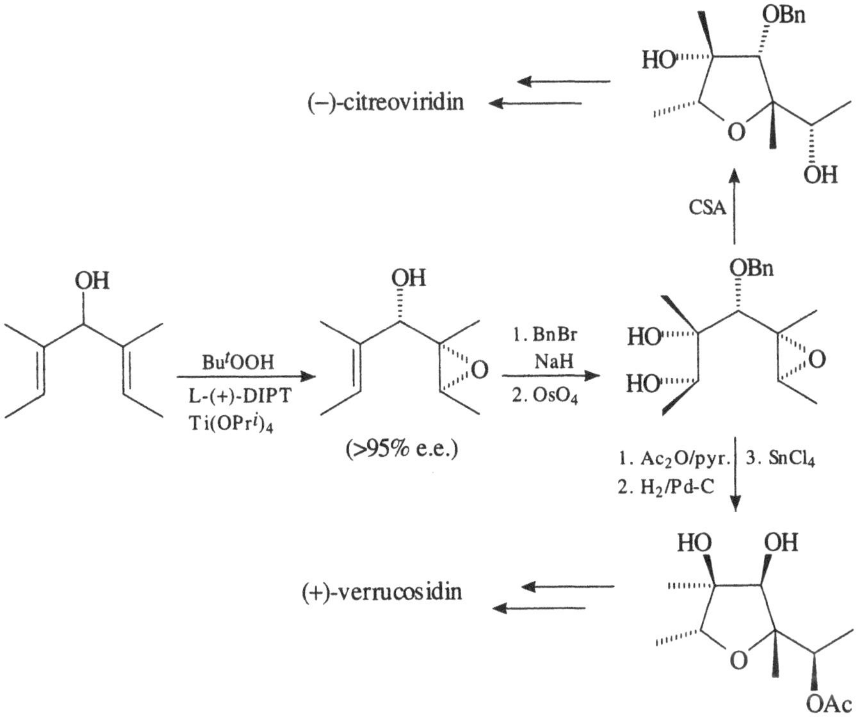

Scheme 6.8

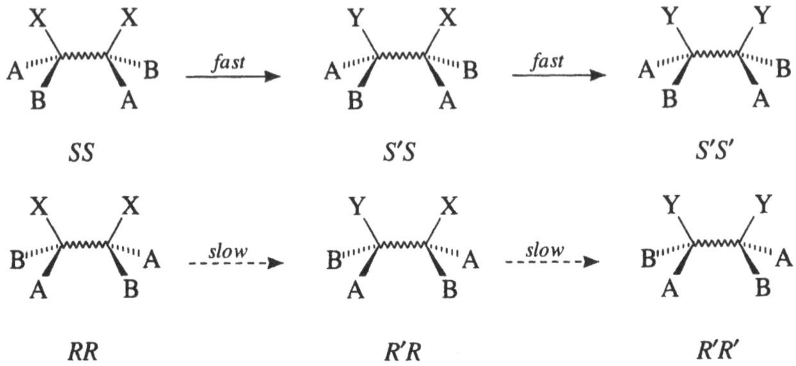

Figure 6.9

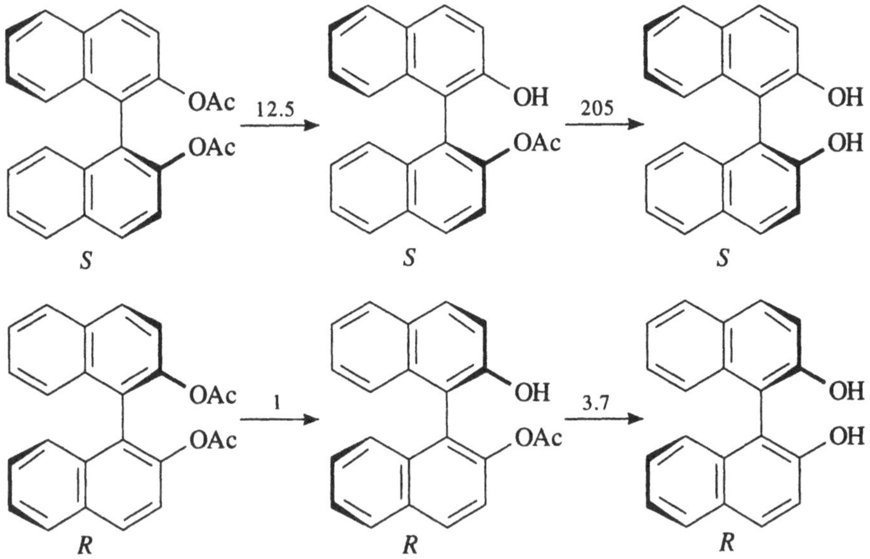

using *Absidia glauca*

Scheme 6.9

(Figure 6.9). An example is shown in Scheme 6.9. Since each enantiomer of the starting material is C_2 symmetric both ester groups in each enantiomer are in identical environments and it is therefore not surprising that if one of the ester

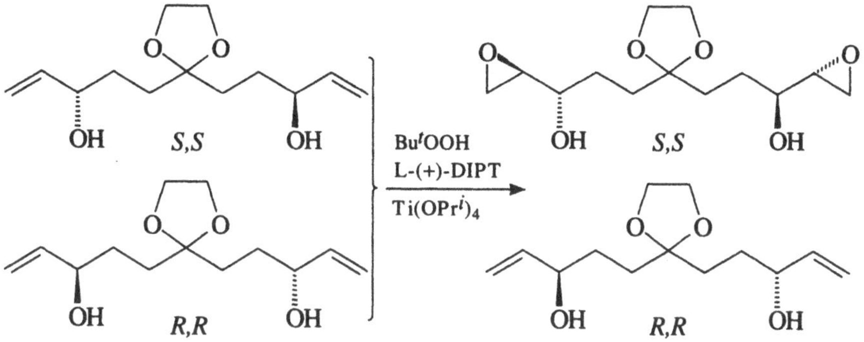

Scheme 6.10

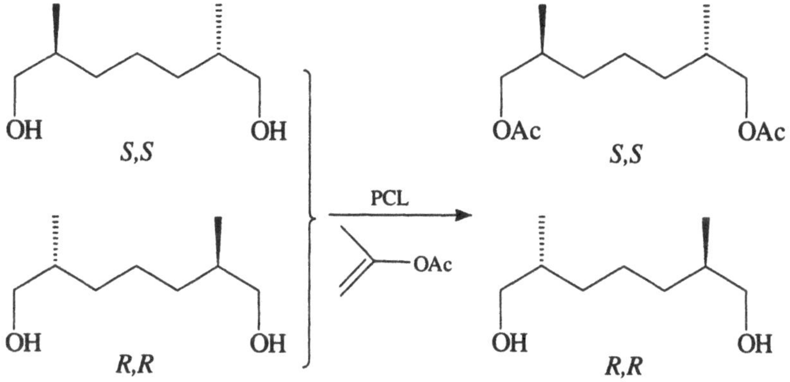

PCL = *Pseudomonas cepacid lipase*

Scheme 6.11

groups in the S enantiomer reacts most rapidly in the first step the other ester group in this isomer will also react more rapidly in the second step. Further examples of this strategy are shown in Schemes 6.10 and 6.11.

PROBLEMS

1. The scheme below shows part of a synthesis of (+)-α-cuparenone. Suggest suitable reagents for each of the steps *1–4*. How could (−)-α-cuparenone be synthesized by a modification of the route shown?

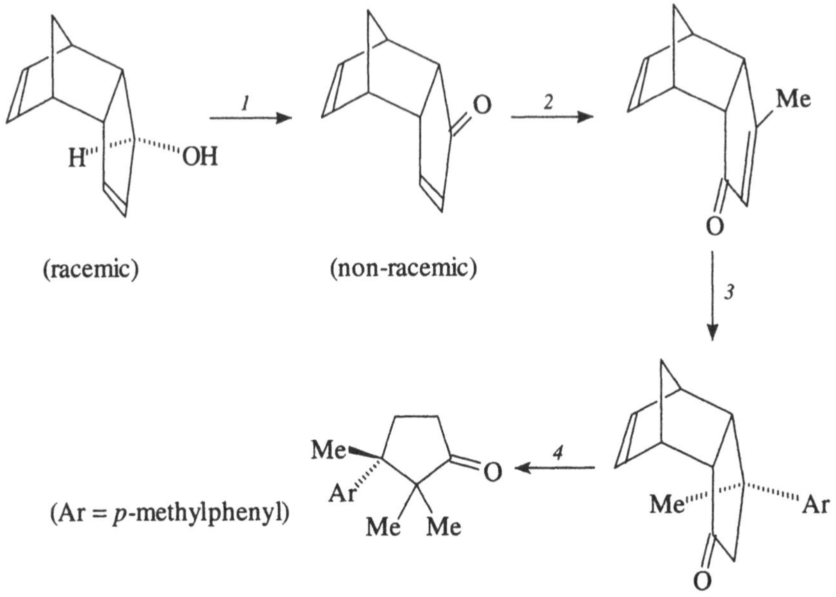

(racemic)   (non-racemic)

(Ar = *p*-methylphenyl)

2. The scheme below shows part of a synthesis of pumiliotoxin C. (i) Suggest
   suitable reagents or a procedure for step *1*; (ii) explain the stereochemical
   outcome of step *4*; and (iii) propose a mechanism for step *6*.

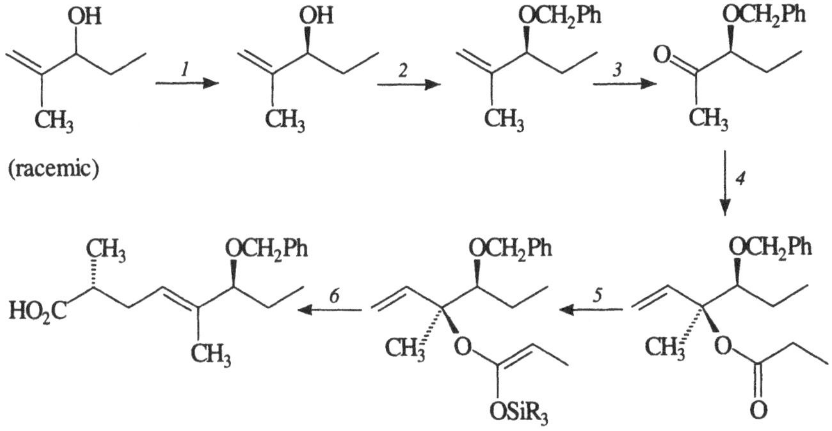

(racemic)

Reagents: *2*, NaH, PhCH₂Br; *3*, O₃; *4*, vinylMgBr then EtCOCl; *5*, LDA,
R₃SiCl, −78°C; *6*, −78°C to room temperature.

3. By considering the *bis*-epoxidation of divinyl carbinol (Scheme 6.4), show that the major products from complete reaction will be a chiral *bis*-epoxide and its achiral stereoisomer.

FURTHER READING

H. B. Kagan and J. C. Fiaud, *Topics in Stereochemistry*, 1988, *18*, 249.
C.-S. Chen, Y. Fujimoto, G. Girdaukas and C. J. Sih, *J. Am. Chem. Soc.*, 1982, *104*, 7294.
Y.-F. Wang, C.-S. Chen, G. Girdaukas and C. J. Sih, *J. Am. Chem. Soc.*, 1984, *106*, 3695.

Asymmetric epoxidation

V. S. Martin, S. S. Woodard, T. Katsumi, Y. Yamada, M. Ikeda and K. B. Sharpless, *J. Am. Chem. Soc.*, 1981, *103*, 6237.
Y. Gao, R. M. Hanson, J. M. Klunder, S. Y. Ko, H. Masamune and K. B. Sharpless, *J. Am. Chem. Soc.*, 1987, *109*, 5765.

Asymmetric dihydroxylation

M. S. VanNieuwenhze and K. B. Sharpless, *J. Am. Chem. Soc.*, 1993, *115*, 7864.
F. J. A. D. Bakkeren, A. J. H. Klunder and B. Zwanenburg, *Tetrahedron*, 1996, *52*, 7901.

Asymmetric hydrogenation

M. Kitamura, I. Kasahara, K. Manabe, R. Noyori and H. Takaya, *J. Org. Chem.*, 1988, *53*, 708.
M. Studer, V. Okafor and H.-U. Blaser, *J. Chem. Soc., Chem. Commun.*, 1998, 1053.

Epoxidation of divinyl carbinols

D. B. Smith, Z. Wang and S. L. Schreiber, *Tetrahedron*, 1990, *46*, 4793.
S. L. Schreiber, T. S. Schreiber and D. B. Smith, *J. Am. Chem. Soc.*, 1987, *109*, 1525.

Sequential kinetic resolution

S.-H. Wu, L.-Q. Zhang, C.-H. Chen, G. Girdaukas and C. J. Sih, *Tetrahedron Lett.*, 1985, *26*, 4323.
Z.-W. Guo, S.-H. Wu, C.-S. Chen, G. Girdaukas and C. J. Sih, *J. Am. Chem. Soc.*, 1990, *112*, 4942.

ANSWERS

1. In principle kinetic resolution could be achieved by Sharpless epoxidation (see answer to question 2), or by enzymic hydrolysis of the corresponding racemic acetate.

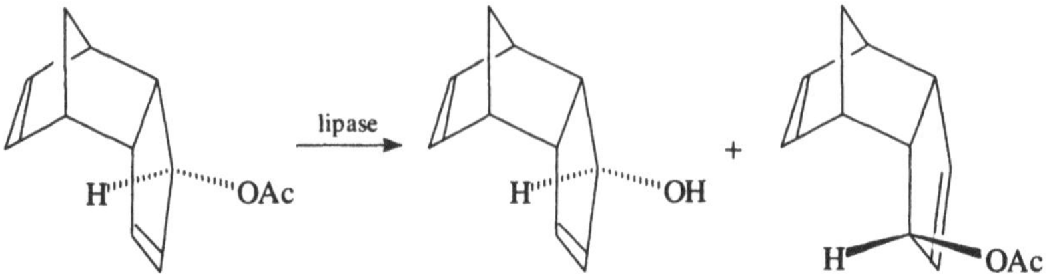
lipase
H
OAc
H
OH
+
H
OAc

(racemic)

The reagents used for the later steps are shown below. Note the stereoselectivity achieved in the conjugate addition step. Clearly the acetate of the other enantiomer obtained from the kinetic resolution could be used to synthesize (−)-α-cuparenone.

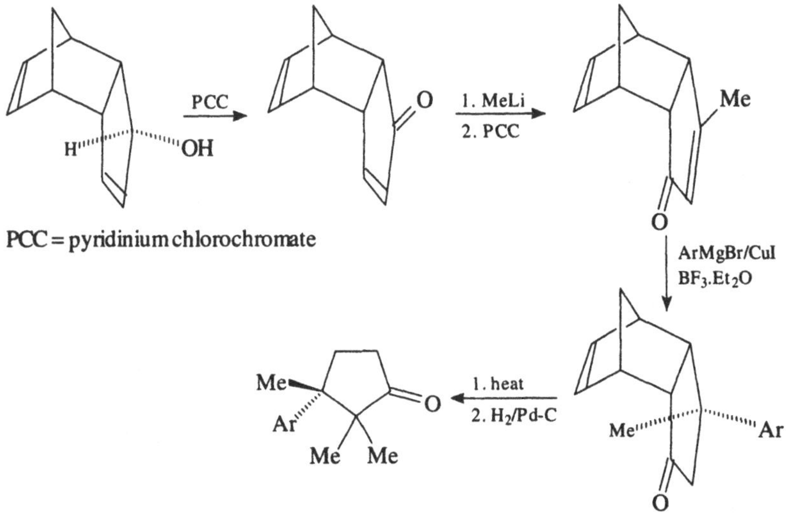
PCC
H
OH
O
1. MeLi
2. PCC
Me
O
PCC = pyridinium chlorochromate
ArMgBr/CuI
BF3.Et2O
Me
Ar
O
Me
Me
1. heat
2. H2/Pd-C
Me
Ar
O

(See S. Tanaka, K. Inoata and K. Ogasawara, *J. Chem. Soc., Chem. Commun.*, 1989, 271)

2. (i)

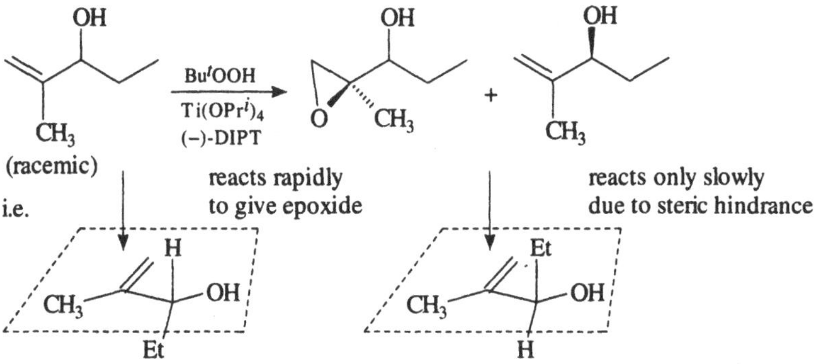
OH
CH3
(racemic)
i.e.
Bu'OOH
Ti(OPr')4
(−)-DIPT
OH
O
CH3
+
OH
CH3
reacts rapidly
to give epoxide
reacts only slowly
due to steric hindrance
H
CH3
OH
Et
Et
CH3
OH
H

or react racemic alcohol with enantiomerically pure acid and separate diastereomeric esters.

(ii) (see Cram's rule, Chapter 2)

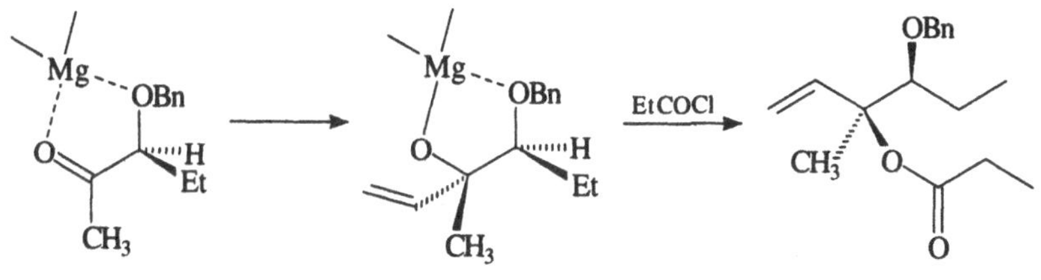

(iii) (see Chapter 3)

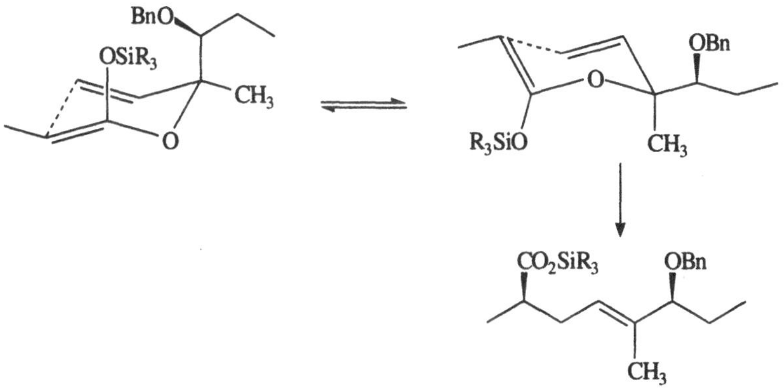

3. Since P_1 is the major product formed in the first step it is logical to assume that the major product(s) from the second step will result from epoxidation of this compound. The diastereoselectivity of the second step will be poor since (+)-DIPT favours epoxidation from the hindered face of P_1 (see Figure 6.8). The two possible products are shown below. It is interesting to note that, of the minor mono-epoxides, P_3 and P_4 would be predicted to react rapidly to give the same two compounds.

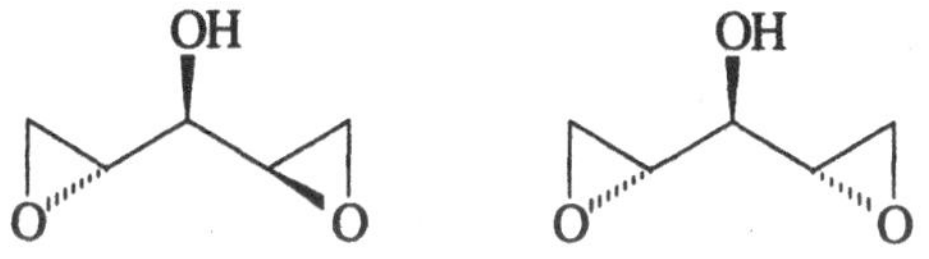

7

Asymmetric Reactions on Molecules with a Plane of Symmetry

When a plane of symmetry is present in a bifunctional molecule the two halves are enantiotopic and can be differentiated by chiral reagents. An enantioselective reaction will convert the molecule into one of two enantiomeric products. Furthermore, in many cases, the enantiomeric purity of the product will increase as the reaction proceeds due to kinetic resolution (cf. Sections 6.3 and 6.4). Some examples are shown in Scheme 7.1.

The reactions shown in Scheme 7.1 all involve enantiotopic selectivity, i.e. preferential attack on one of two (or more) identical atoms or groups in a prochiral molecule. These particular examples also involve diastereofacial selectivity, i.e. preferential attack on one face of a double bond in a molecule which contains one (or more) chiral and/or prochiral centres. Several examples of the asymmetric epoxidation reaction of compounds with a plane of symmetry were included in Section 6.4. Two further examples involving *meso* compounds are shown in eqns 7.1 and 7.2. The same principles apply to the asymmetric hydroboration reactions of symmetrical cyclopentadienes and the ene reactions of bicyclo[3.3.0]octa-2,7-diene (Scheme 7.1). A further example of the stereoselective hydroboration of a *meso* compound is shown in eqn. 7.3.

These reactions are examples of transformations in which bifunctional compounds having a plane of symmetry, including *meso* compounds, are converted into chiral products. There are many examples of enzyme-catalysed reactions of this type (e.g. eqns 7.4 and 7.5). In addition there are an ever increasing number of non-enzymatic examples. In all such reactions the enantiomeric excess of the product is enhanced by kinetic resolution (Figure 7.1 and cf. Section 6.2).

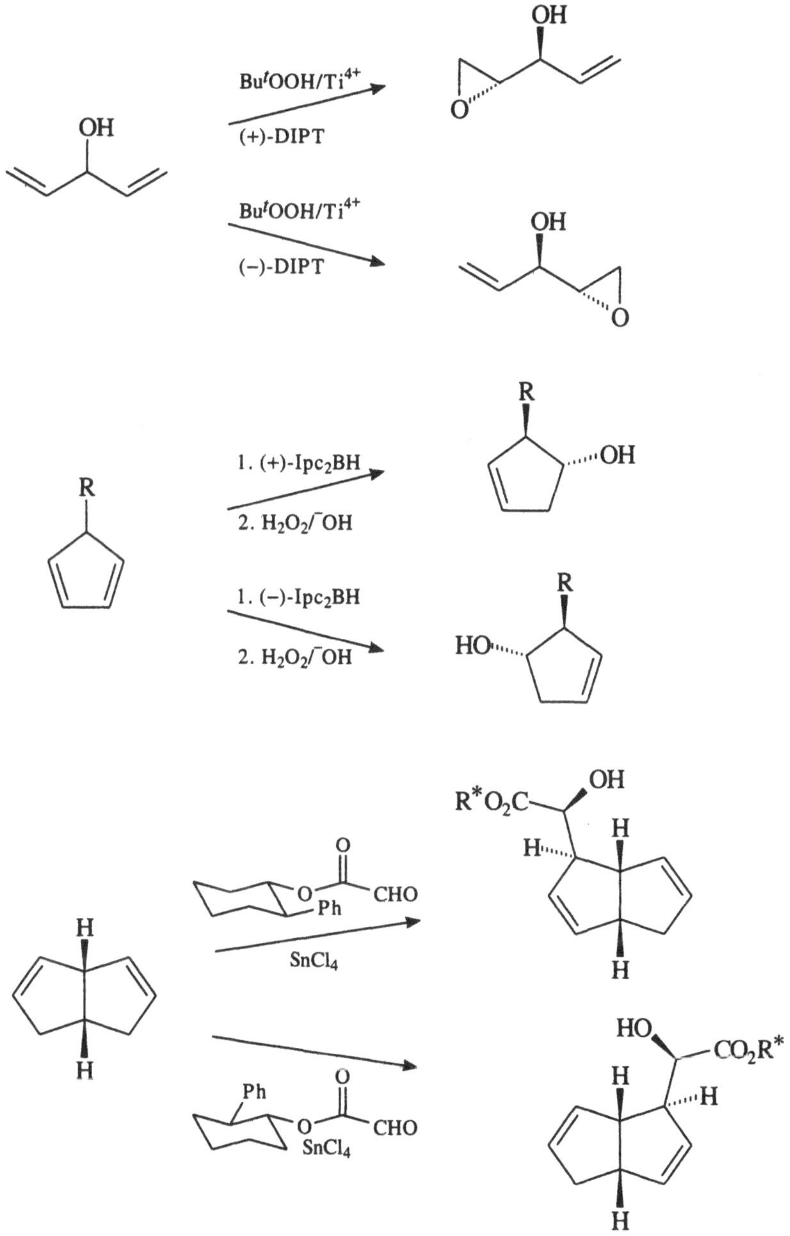
OH
Bu^{t}OOH/Ti^{4+}
(+)-DIPT
OH
O
Bu^{t}OOH/Ti^{4+}
(−)-DIPT
OH
O
R
1. (+)-Ipc$_2$BH
2. H$_2$O$_2$/$^-$OH
R
OH
1. (−)-Ipc$_2$BH
2. H$_2$O$_2$/$^-$OH
R
HO
H
H
O
O
CHO
Ph
SnCl$_4$
R*O$_2$C
OH
H
H
H
Ph
O
O
CHO
SnCl$_4$
HO
CO$_2$R*
H
H
H

Scheme 7.1

$$(7.1)$$

>99% e.e.

$$(7.2)$$

$$(7.3)$$

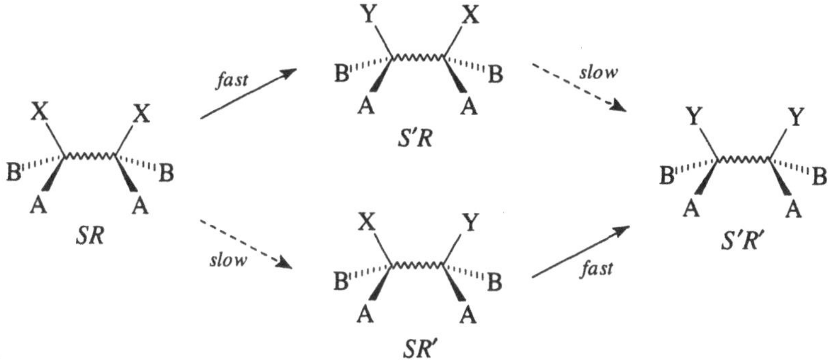

Figure 7.1

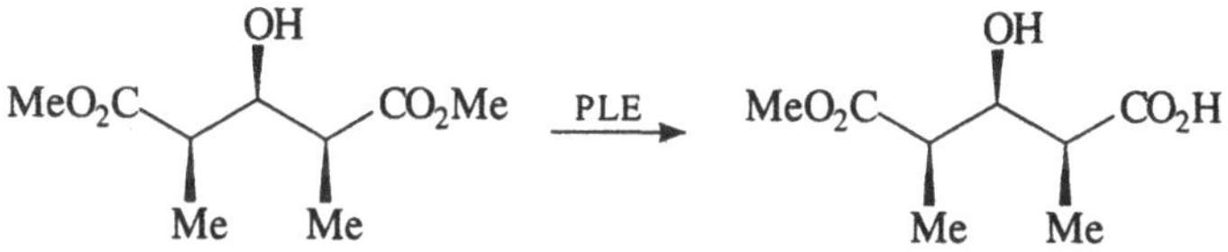

$$MeO_2C \overset{OH}{\wedge} CO_2Me \xrightarrow{PLE} MeO_2C \overset{OH}{\wedge} CO_2H \quad 98\% \text{ e.e.}$$

PLE = *pig liver esterase*

(7.4)

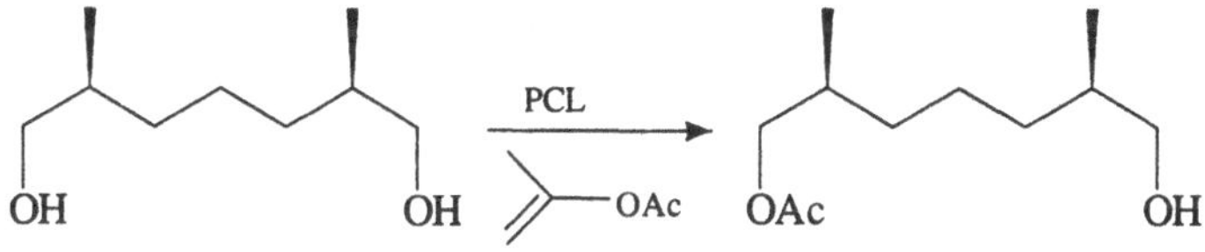

$$\xrightarrow{PCL} \quad 95\% \text{ e.e.}$$

PCL = *Pseudomonas cepacid lipase*

(7.5)

The symmetry of a *meso* compound can also be broken by reaction of both functional groups with a chiral reagent. This approach is discussed in Section 7.9.

7.1 ENANTIOSELECTIVE OPENING OF EPOXIDES

Several chiral lithium amide bases have been developed which are capable of selectively abstracting enantiotopic protons from organic compounds. They can be used to convert symmetrical epoxides into enantiomerically enriched allylic alcohols (eqns 7.6 and 7.7).

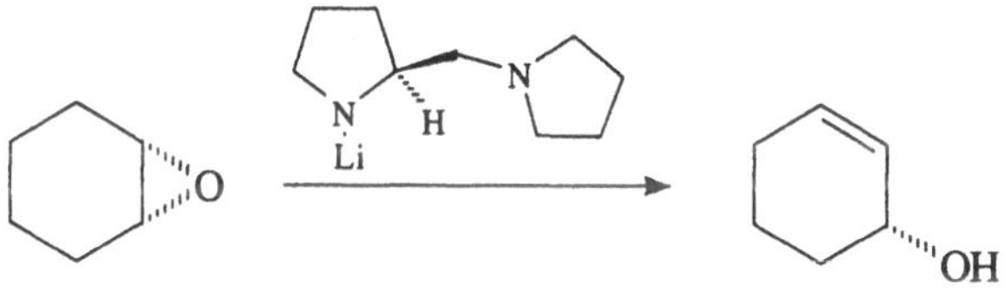

92% e.e. (7.6)

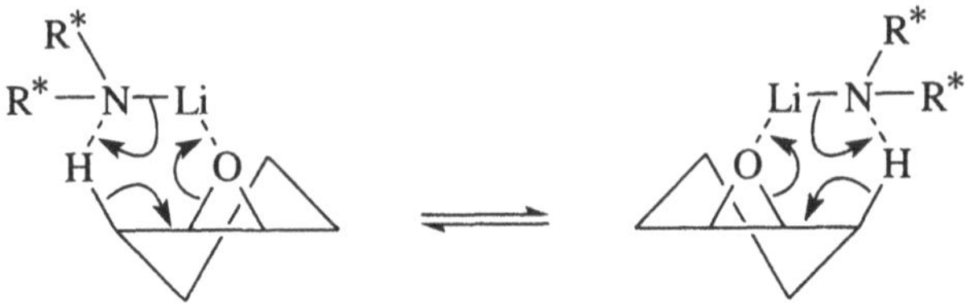

Figure 7.2

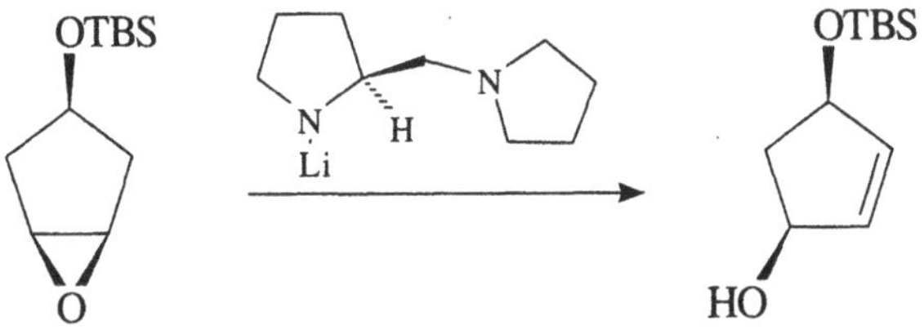

90% e.e.

(7.7)

Since the deprotonation of cyclohexene oxide (eqn. 7.6) involves abstraction of a quasi-axial proton, the observed enantioselectivity is due to differentiation between the two equilibrating conformations (Figure 7.2).

An alternative approach involves reacting the epoxide with a nucleophile in the presence of a chiral Lewis acid (eqn. 7.8).

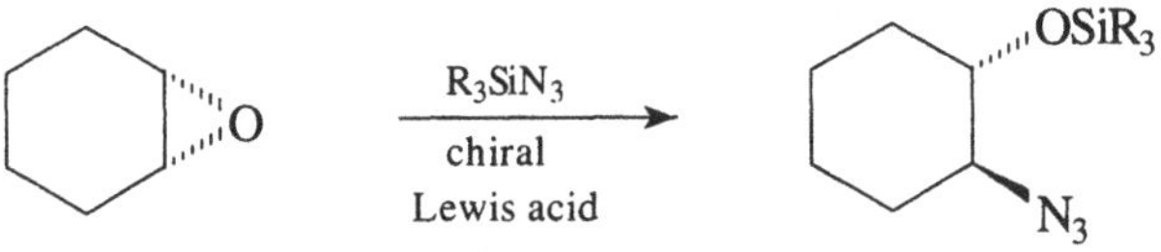

93% e.e.

(7.8)

7.2 ENANTIOSELECTIVE DEPROTONATION OF KETONES

Chiral bases can also be used to convert prochiral ketones into non-racemic silyl enol ethers which can then be converted into other chiral target molecules (eqns 7.9 and 7.10).

7.3 HORNER–EMMONS REACTION

In a similar manner chiral non-racemic phosphonamides and phosphonate esters react with prochiral ketones to afford chiral alkenes (Scheme 7.2).

Prochiral aldehydes can also be utilized (eqn. 7.11). In this case, reaction of a dialdehyde with one equivalent of a chiral phosphonate leads to a chiral product. Removal of the chiral auxiliary group would yield an $\alpha\beta$-unsaturated acid containing three stereogenic centres.

$$(7.9)$$

$$(7.10)$$

carbacyclin

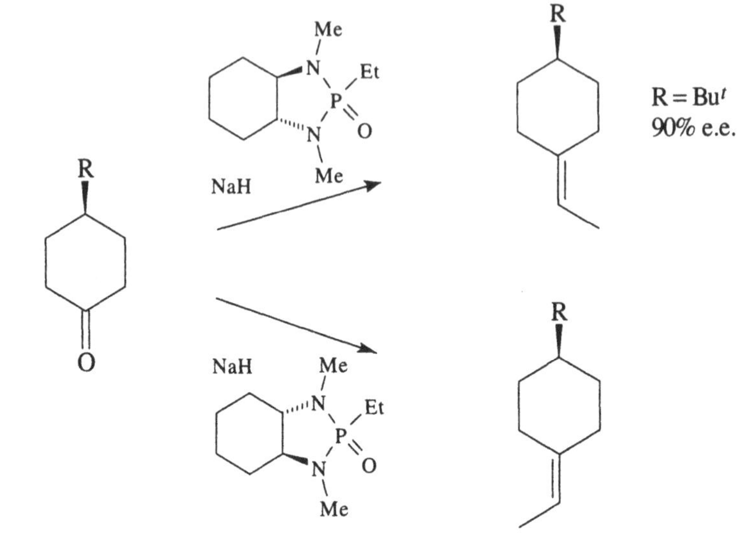

Scheme 7.2

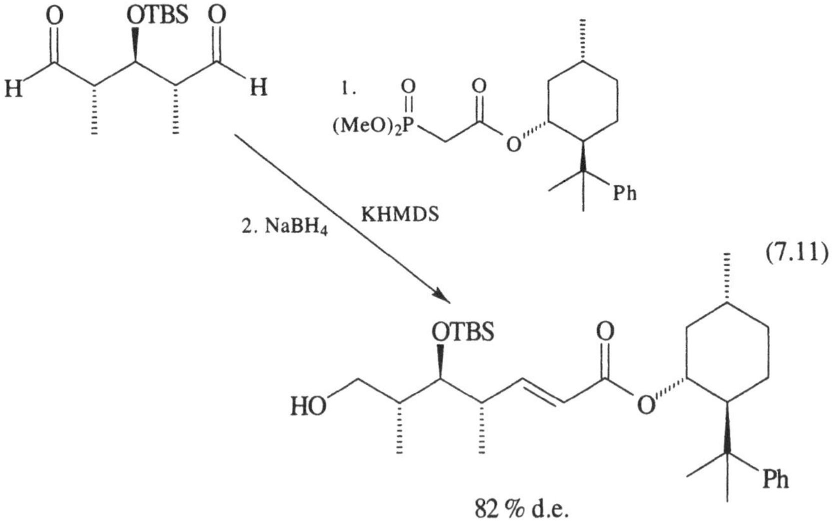

(7.11)

82 % d.e.

7.4 STEREOSELECTIVE REACTIONS ON CYCLIC ANHYDRIDES

Cyclic anhydrides afford versatile bifunctional starting materials for organic synthesis. They are particularly useful for the synthesis of monoesters and lactones (eqns 7.12 and 7.13 and Scheme 7.3).

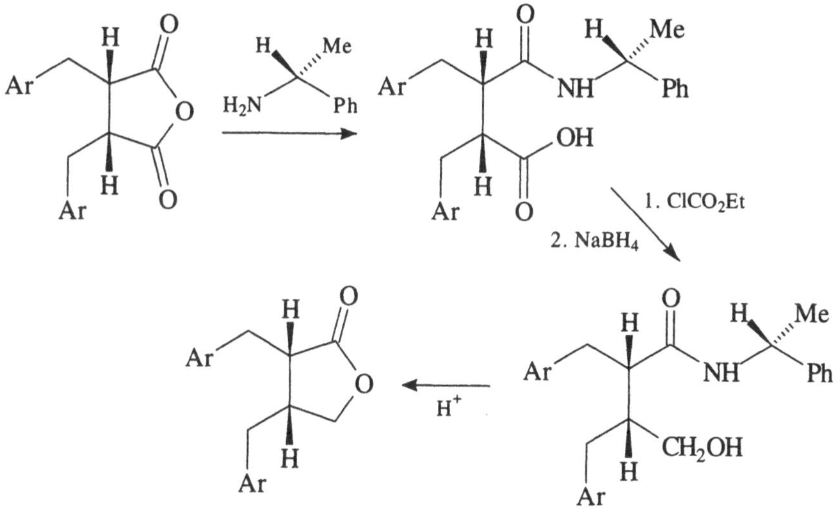

Scheme 7.3

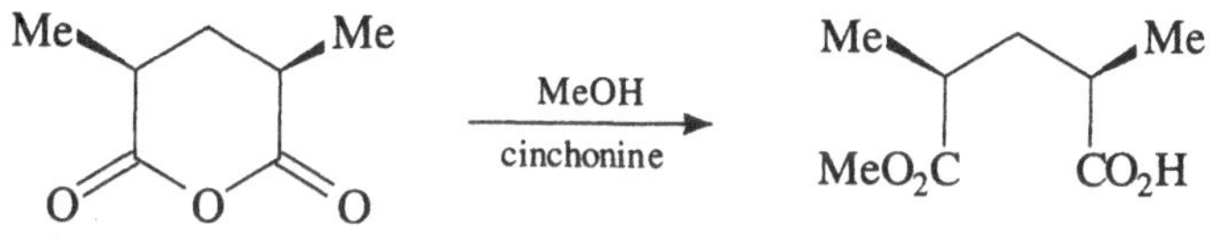

(S,R) 70% e.e.

(7.12)

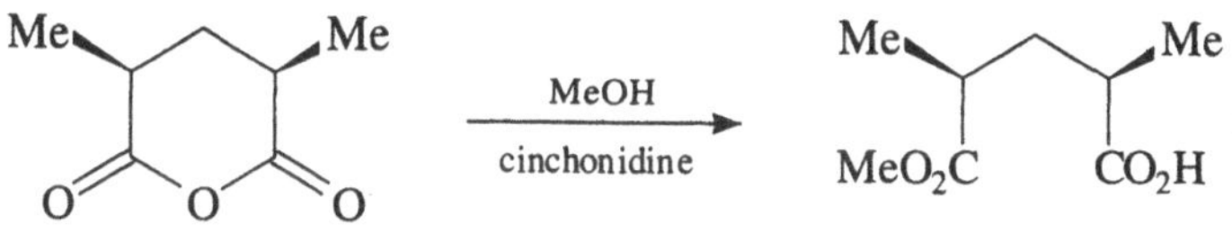

(R,S) 64% e.e.

(7.13)

7.5 STEREOSELECTIVE REACTIONS OF DIENES

Several examples of asymmetric epoxidation and hydroboration have already been discussed. An example involving asymmetric dihydroxylation is shown in Scheme 7.4.

7.6 STEREOSELECTIVE REACTIONS OF DIOLS

In addition to enzyme-catalysed monoesterification (eqn. 7.5), diols can be "desymmetrized" by reaction (of their trimethylsilyl ethers) with a chiral non-racemic ketone (eqn. 7.14 and Scheme 7.5).

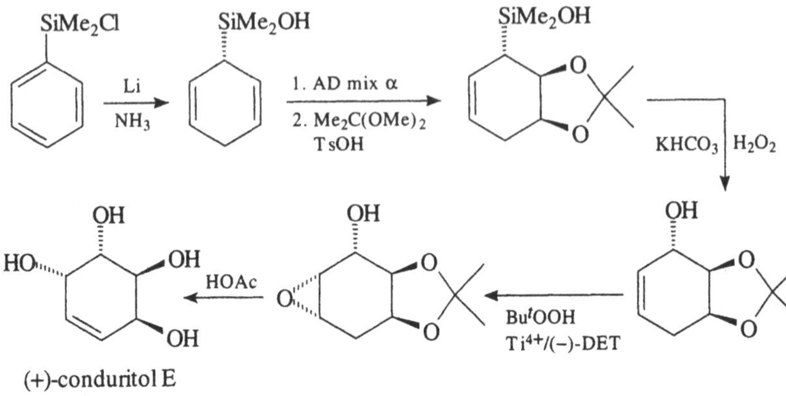

Scheme 7.4

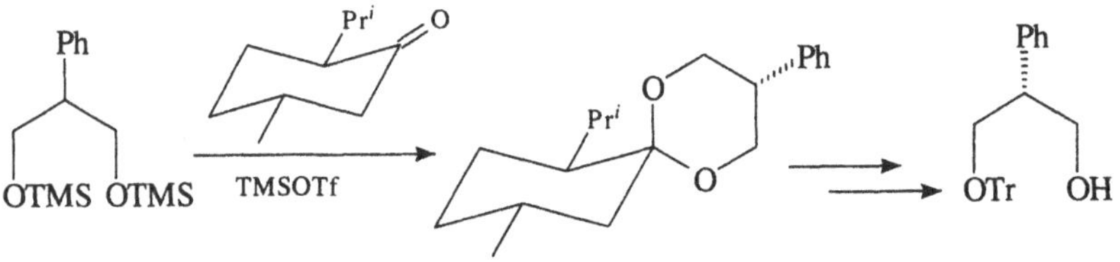

major diastereomer
89% d.e.

(7.14)

A second approach involving reaction of glycerol with the enol ether of a chiral diketone is shown in eqn. 7.15. Notice that it is the secondary OH group and one of the enantiotopic primary OH groups which react to form the spiro-ketal. Benzylation of the remaining OH group followed by removal of the chiral protecting group give the chiral non-racemic monobenzyl ether.

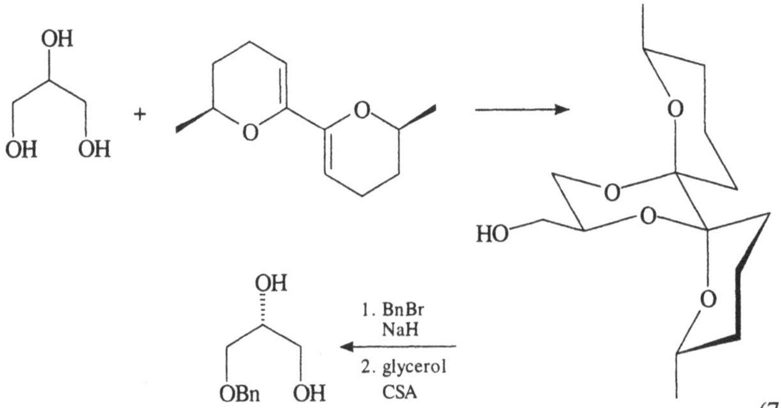

(7.15)

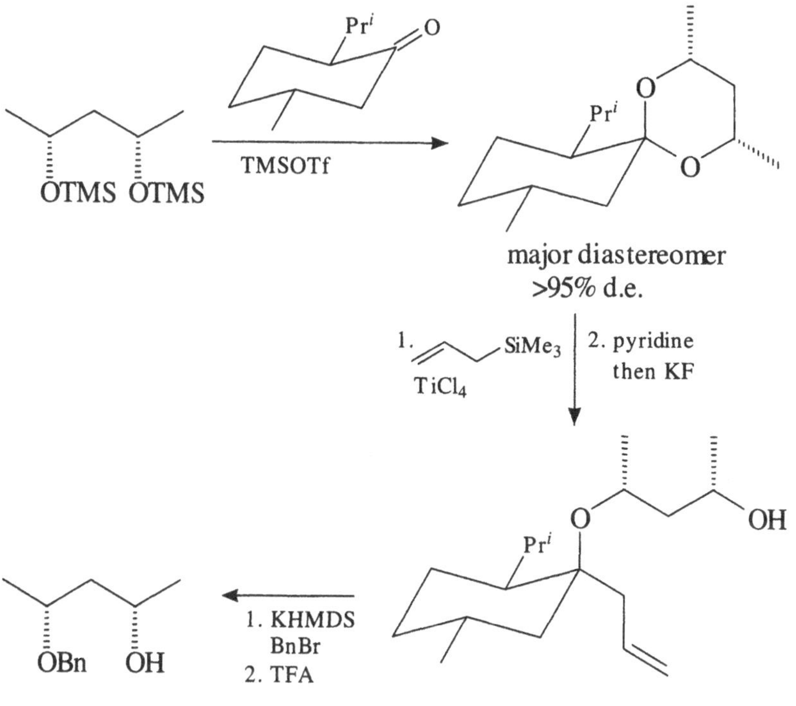

Scheme 7.5

7.7 ASYMMETRIC REACTIONS ON DIKETONES, DIACIDS AND DILACTONES

An elegant cyclization of a symmetrical triketone is involved in eqn. 7.16, and the construction of a stereoselectively protected cyclopentadienone is shown in eqn. 7.17. In the latter case conjugate addition to the $\alpha\beta$-unsaturated ketone followed by removal of the cyclopentadiene protecting group would give a chiral non-racemic cyclopentenone.

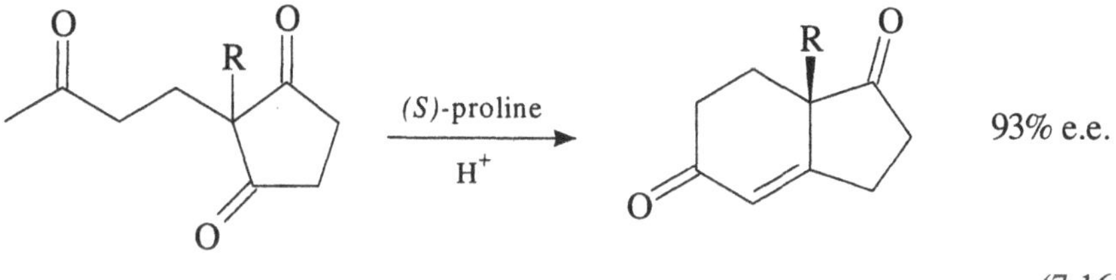

93% e.e.

(7.16)

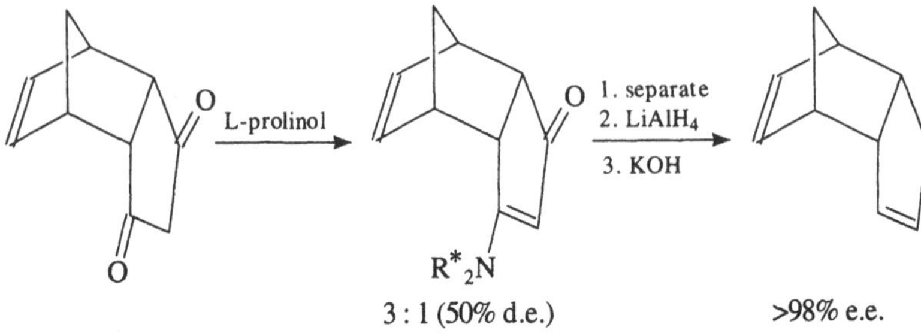

3:1 (50% d.e.)   >98% e.e.   (7.17)

Similarly the cyclization of a symmetrical diacid to give an achiral monolactone is shown in eqn. 7.18, and the ring opening of a symmetrical dilactone is shown in eqn. 7.19.

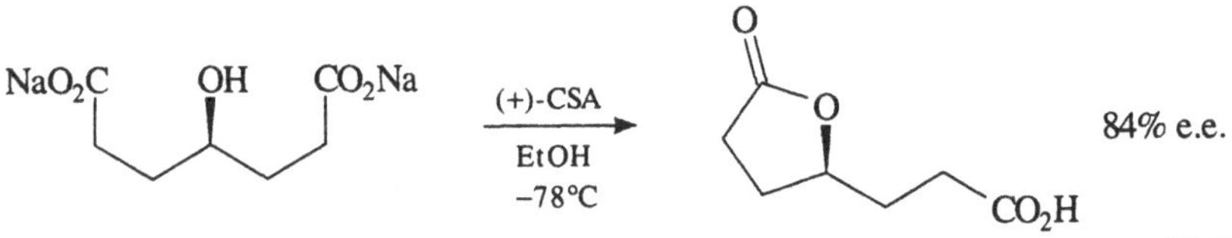

84% e.e.

(7.18)

up to 68% d.e.

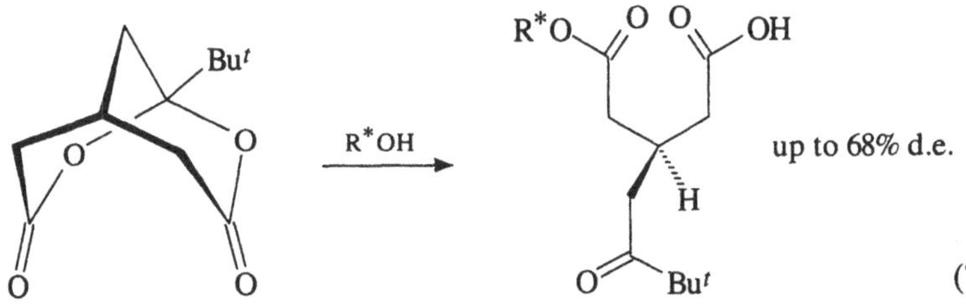

(7.19)

7.8 ELIMINATION REACTIONS

By converting a racemic alkene into a prochiral product by addition of HCl it is possible to use a chiral base to bring about stereoselective elimination and hence deracemization (eqn. 7.20).

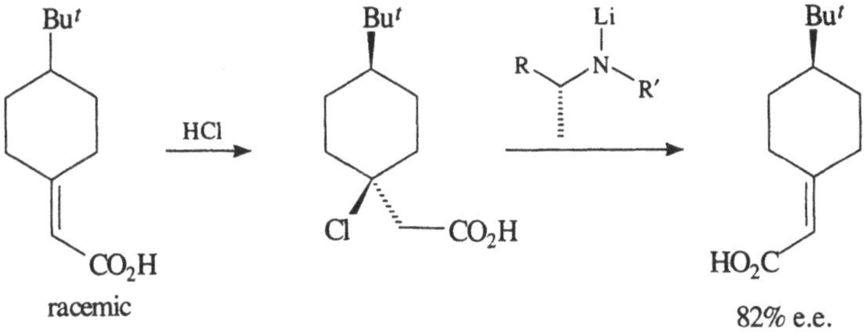

racemic

82% e.e.   (7.20)

7.9 TERMINUS DIFFERENTIATION BY REACTION AT BOTH GROUPS

An alternative method to convert a compound with a plane of symmetry into a chiral non-racemic product is to react both enantiotopic groups with a chiral reagent. For example, reaction of a diacid (or its anhydride) with two equivalents of a chiral auxiliary gives a chiral non-racemic product which reacts with other achiral reagents in a stereoselective manner (Schemes 7.6 and 7.7).

In the sequence shown on page 125 reaction of (−)-diisopinocampheylallylborane with a dialdehyde introduces two new stereogenic centres and fixes the configuration at five existing centres (Scheme 7.8). The two termini are then readily differentiated by conversion to the 1,3-acetonide. The diastereotopic group selectivity exhibited in this step can be attributed to the thermodynamic preference for the *syn*-1,3-acetonide rather than the *anti*-1,3-acetonide due to the 1,3-diaxial interaction present in the *anti*-isomer.

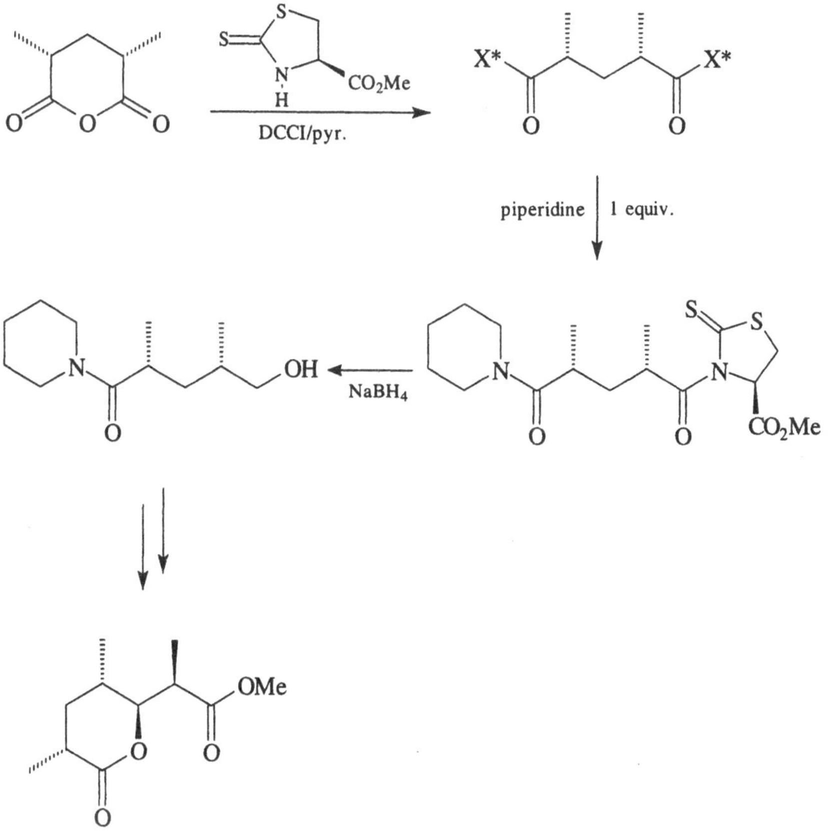

Scheme 7.6

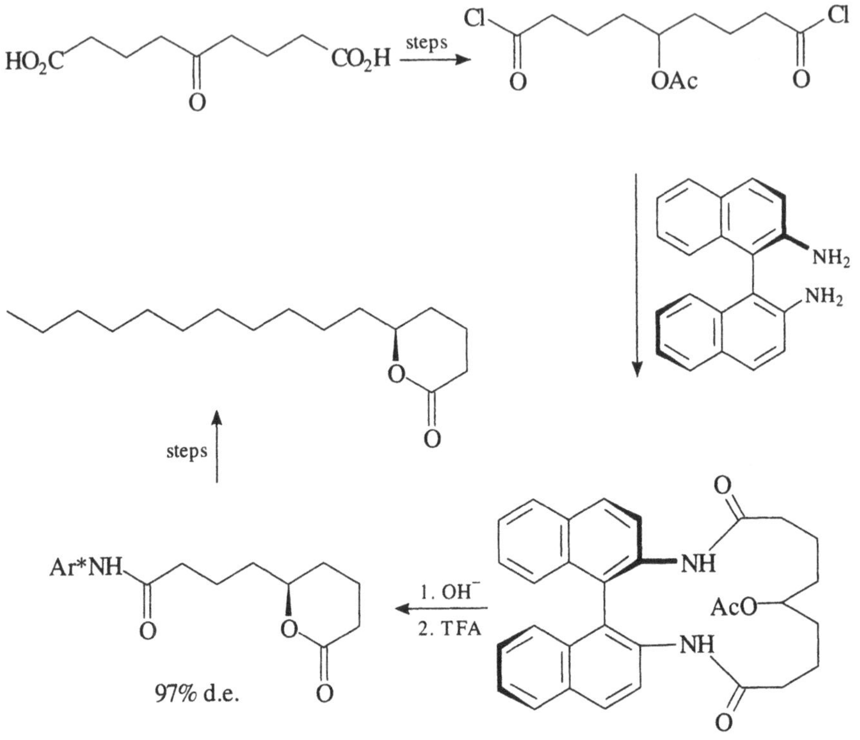

Scheme 7.7

Another example of this approach is involved in the synthesis of a precursor of the halichondrins (Scheme 7.9). Here epoxidation of both allylic alcohol groups yields a chiral non-racemic product containing six stereogenic centres. Stereoselective opening of both epoxides followed by lactone formation and [3,3] sigmatropic rearrangement then proceeds under substrate control without the need for other control elements or chiral reagents.

A second example involving the Sharpless epoxidation is provided by the synthesis of the 2,5-linked *bis*-tetrahydrofurans (Scheme 7.10, cf. Scheme 8.17). In this case the asymmetric epoxidation creates four new stereogenic centres, the absolute configuration of which is dictated by the configuration of the tartrate ester used. Removal of the ketal protecting group by acid hydrolysis releases the epoxy-alcohol units which react stereoselectively to give the enantiomeric *bis*-tetrahydrofuran isomers.

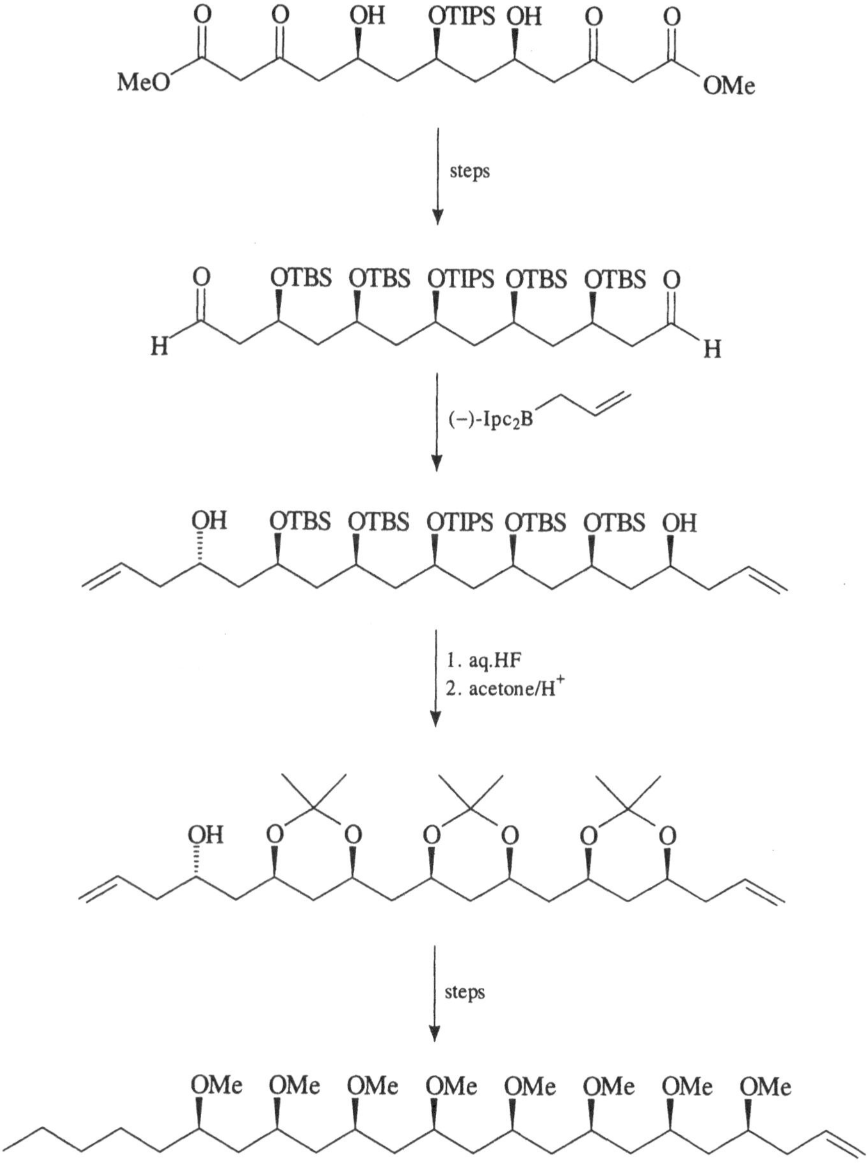
O
O
OH
OTIPS
OH
O
O
MeO
OMe
steps
O
OTBS
OTBS
OTIPS
OTBS
OTBS
O
H
H
(−)-Ipc₂B
OH
OTBS
OTBS
OTIPS
OTBS
OTBS
OH
1. aq.HF
2. acetone/H⁺
OH
O
O
O
O
O
O
steps
OMe
OMe
OMe
OMe
OMe
OMe
OMe
OMe

Scheme 7.8

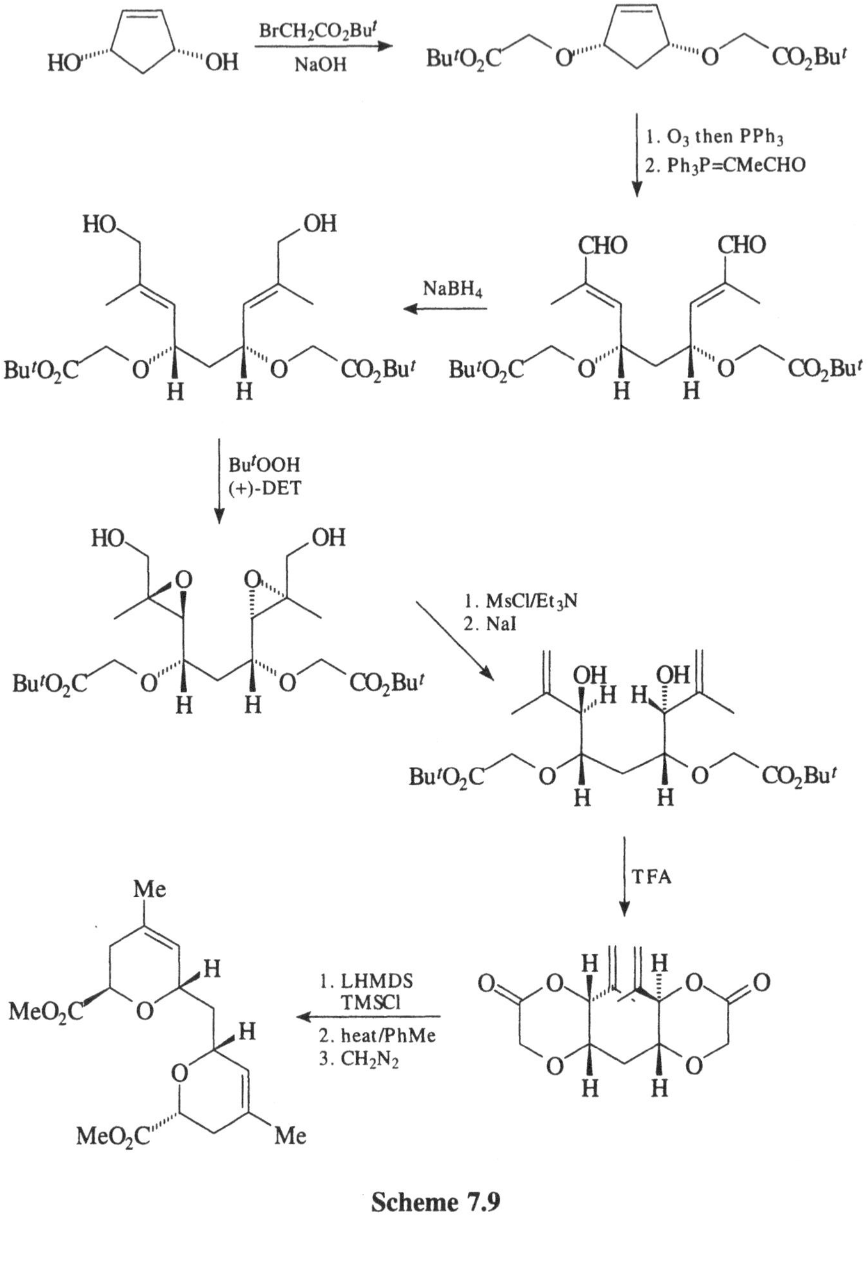
HO
OH
$BrCH_2CO_2Bu^t$
NaOH
Bu^tO_2C
O
O
CO_2Bu^t
1. O_3 then PPh_3
2. $Ph_3P{=}CMeCHO$
CHO
CHO
$NaBH_4$
HO
OH
H
H
Bu^tOOH
(+)-DET
1. MsCl/Et_3N
2. NaI
OH
OH
TFA
1. LHMDS
TMSCl
2. heat/PhMe
3. CH_2N_2
Me
MeO_2C
O
H

Scheme 7.9

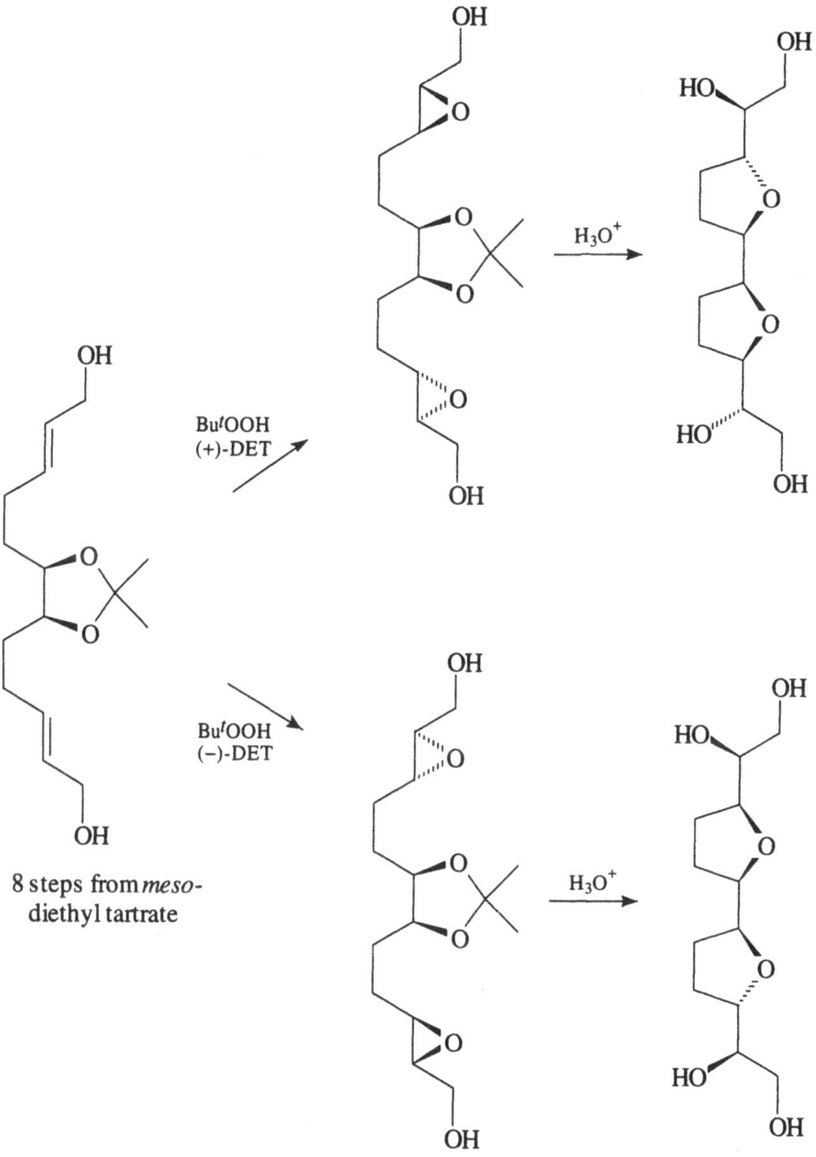

OH
OH
HO
O
O
O
H3O+
O
OH
O
Bu^tOOH
(+)-DET
HO
OH
OH
O
O
OH
OH
Bu^tOOH
(−)-DET
OH
HO
O
OH
8 steps from *meso*-
diethyl tartrate
O
H3O+
O
O
O
HO
OH
OH

Scheme 7.10

PROBLEMS

1. The scheme below shows the synthesis of a non-racemic chiral lactone **C** from an achiral diol **A**. (i) Suggest a method to convert **A** into **B** (THP = tetrahydropyranyl protecting group). (ii) Suggest a mechanism for the conversion of **B** into **C**. (iii) How could the enantiomer of **B** be converted into **C**? (iv) How could **A** be converted into the enantiomer of **C**?

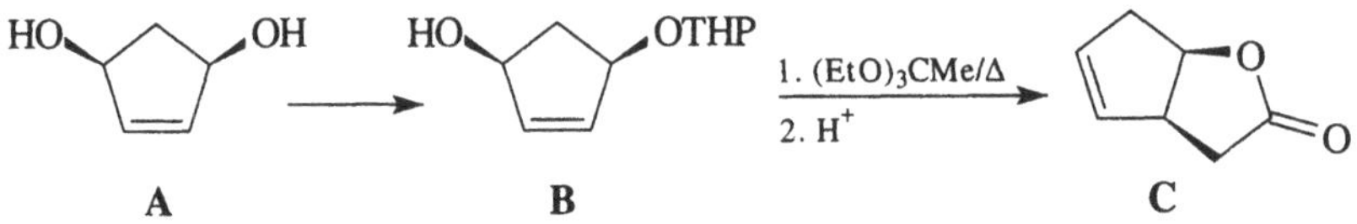

$$\text{A} \qquad \text{B} \qquad \text{C}$$

Reagents over arrow: 1. $(EtO)_3CMe/\Delta$ 2. H^+

2. The symmetrical ketone **D** can be readily prepared as shown in the following scheme. Suggest a method by which **D** could be converted into a non-racemic chiral product, specifically with a view to preparing a chiral acyclic product such as **E**.

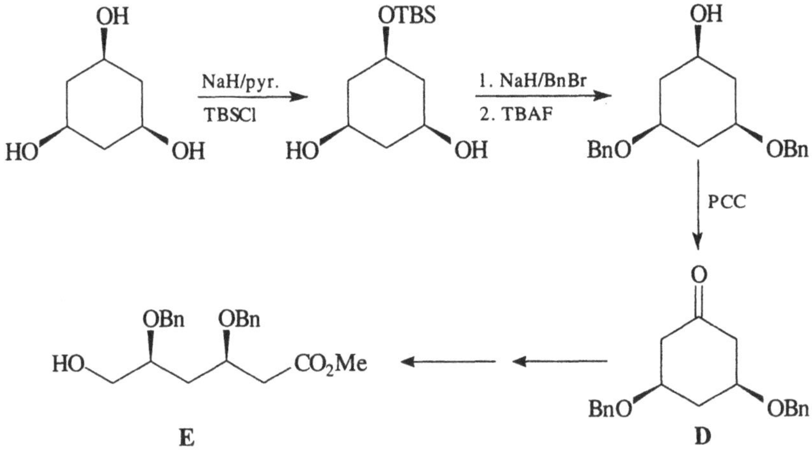

Reagents: NaH/pyr., TBSCl ; 1. NaH/BnBr, 2. TBAF ; PCC

$$\text{E} \qquad \text{D}$$

FURTHER READING

R. S. Ward, *Chem. Soc. Rev.*, 1990, *19*, 1.
C. S. Poss and S. L. Schreiber, *Acc. Chem. Res.*, 1994, *27*, 9.
S. R. Magnuson, *Tetrahedron*, 1995, *51*, 2167.

Reactions of epoxides

D. M. Hodgson and G. P. Lee, *J. Chem. Soc., Chem. Commun.*, 1996, 1015.
D. M. Hodgson, A. R. Gibbs and G. P. Lee, *Tetrahedron*, 1996, *52*, 14361.
N. Haddad, M. Grishko and A. Brik, *Tetrahedron Lett.*, 1997, *38*, 6075.
W. A. Nugent, *J. Am. Chem. Soc.*, 1992, *114*, 2768.

Deprotonation of ketones

N. S. Simpkins, *Chem. Soc. Rev.*, 1990, *19*, 335.

C. M. Cain, R. P. C. Cousins, G. Coumbarides and N. S. Simpkins, *Tetrahedron*, 1990, *46*, 523.

J. Leonard, J. D. Hewitt, D. Ouali, S. K. Rahman, S. J. Simpson and R. F. Newton, *Tetrahedron Asymmetry*, 1990, *1*, 699.

H. Waldemann, in *Organic Synthesis Highlights II* (ed. H. Waldemann), VCH Publ., Weinheim, 1995, p. 19.

E. G. Occhiato, D. Scarpi, G. Menchi and A. Guarna, *Tetrahedron Asymmetry*, 1996, *7*, 1929.

P O'Brien, *J. Chem. Soc., Perkin Trans. 1*, 1998, 1439.

Horner-Emmons reactions

S. Hanessian, D. Delorme, S. Beaudoin and Y. Leblanc, *J. Am. Chem. Soc.*, 1984, *106*, 5754.

Reactions of anhydrides

S. J. Hulme, P. R. Jenkins, J. Fawcette and D. R. Russell, *Tetrahedron Lett.*, 1994, *35*, 5501.

D. Seebach, G. Jaeschke and Y.-M. Wang, *Angew. Chem. Int. Ed.*, 1995, *34*, 2395.

R. S. Ward, A. Pelter, M. I. Edwards and J. Gilmore, *Tetrahedron*, 1996, *52*, 12799.

R. Verma and S. K. Ghosh, *J. Chem. Soc., Chem. Commun.*, 1996, 1601.

S. Sano, H. Ushirogochi, K. Morimoto, S. Tamai and Y. Nagao, *J. Chem. Soc., Chem. Commun.*, 1996, 1775.

Reactions of dienes

K. Fuji, M. Node, Y. Naniwa and T. Kawabata, *Tetrahedron Lett.*, 1990, *31*, 3175.

S. D. Burke, J. L. Buchanan and J. D. Rovin, *Tetrahedron Lett.*, 1991, *32*, 3961.

R. Angelaud and Y. Landais, *J. Org. Chem.*, 1996, *61*, 5202.

Z. Wang and D. Deschenes, *J. Am. Chem. Soc.*, 1992, *114*, 1090.

Reactions of diols

R. Chenevert and M. Desjardins, *J. Org. Chem.*, 1996, *61*, 1219.

ANSWERS

1. (i)

A (R* = chiral auxiliary group)

(ii)

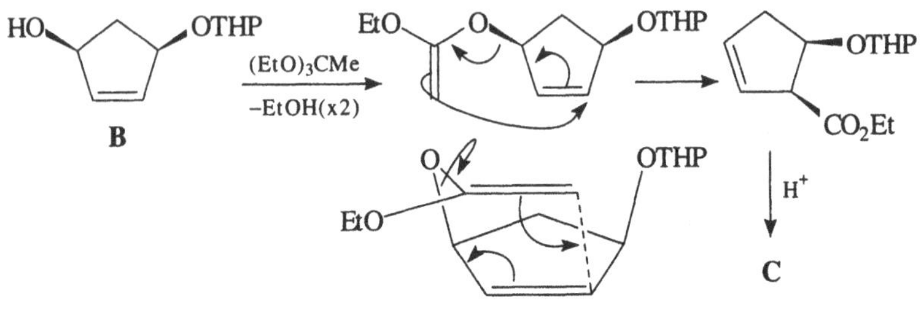

(iii)

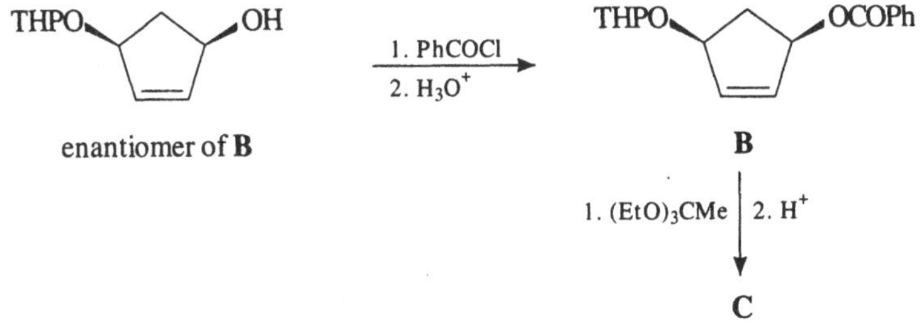

enantiomer of **B**

1. PhCOCl
2. H_3O^+

B

1. $(EtO)_3CMe$ | 2. H^+

C

(iv) Use enantiomer of R*COCl, or use PhCOCl then $(EtO)_3CMe$ on **B**, or use $(EtO)_3CMe$ directly on enantiomer of **B**. (see M. Nara, S. Terashima and S. Yamada, *Tetrahedron*, 1980, *36*, 3161 and 3171).

2. The transformation was performed using a chiral base and trimethylsilyl chloride to form the chiral silyl enol ether **F** (cf. Section 7.2) which was then ozonized and converted into **E** (see T. Honda, S. Ono, H. Mizutani and K. O. Halliman, *Tetrahedron Asymmetry*, 1997, *8*, 181). An alternative strategy might be to carry out a Baeyer–Villiger oxidation using a chiral peracid.

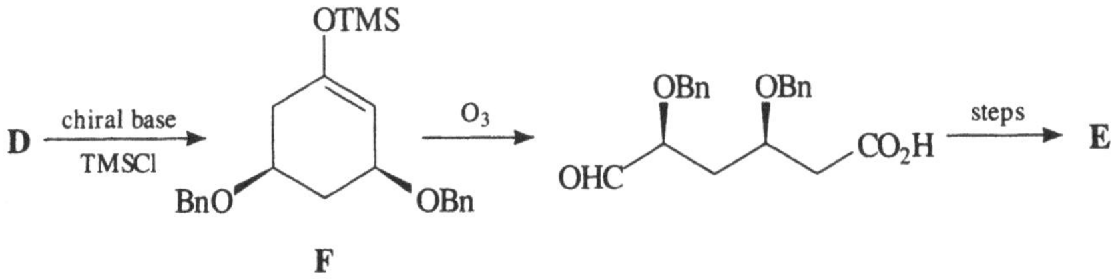

F

8

Asymmetric Reactions on Molecules with C₂ Symmetry

Just as molecules having a plane of symmetry, such as *meso* compounds, can be elegantly and efficiently converted into chiral non-racemic products, similarly molecules having C_2 symmetry provide valuable starting materials for the stereoselective synthesis of polyfunctional compounds. Readily available compounds which possess C_2 symmetry can be used as starting materials for the synthesis of other molecules with C_2 symmetry (Section 8.1), and can also be converted into compounds lacking C_2 symmetry (Section 8.2).

8.1 SYNTHESES OF MOLECULES WITH C₂ SYMMETRY

A sequence which illustrates the synthesis of a functionalised ionophore having C_2 symmetry starting from (+)-diethyl tartrate is shown overleaf (Scheme 8.1). Since the starting material is chiral all of the reactions in which new stereogenic centres are created proceed under substrate control. For example, hydroboration using 9-borabicyclononane proceeds stereoselectively to create two new stereogenic centres whose configuration is determined by the chiral centres already present. Similarly the palladium acetate catalysed cyclization of the diol proceeds stereoselectively with the creation of two further stereogenic centres.

Two further syntheses of compounds having C_2 symmetry are shown below. Scheme 8.2 shows the asymmetric synthesis of two C_2 symmetric *bis*-epoxides from D-mannitol. Scheme 8.3 shows the synthesis of a racemic *bis*-ketal. In this case the C_2 symmetry is created by diastereoselective reduction of a β-hydroxy-ketone using tetramethylammonium triacetoxy-borohydride. This is followed by a substrate controlled hydroboration using 9-BBN.

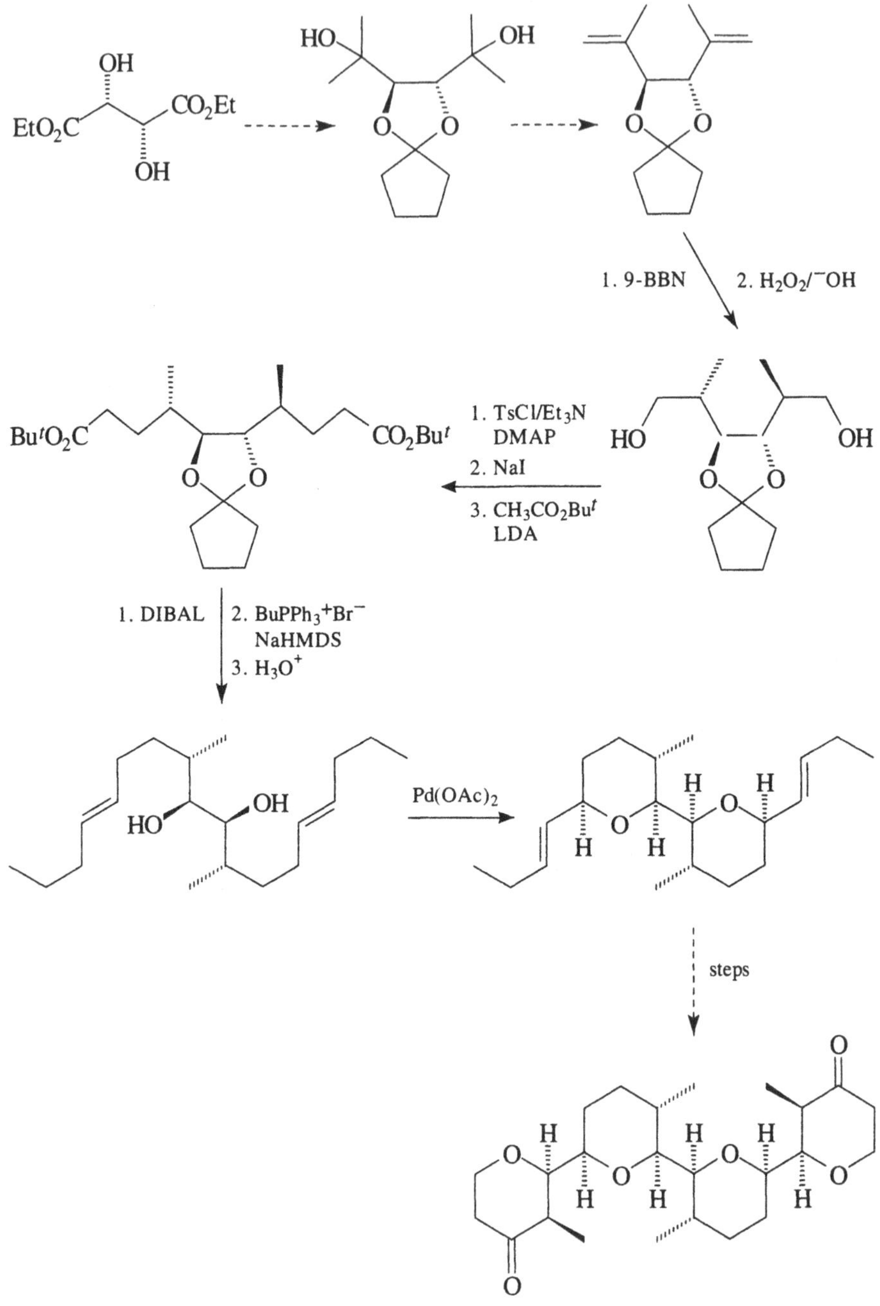
OH
EtO_2C
CO_2Et
OH
HO
OH
O
O
1. 9-BBN
2. $H_2O_2/^{-}OH$
Bu^tO_2C
CO_2Bu^t
1. TsCl/Et_3N DMAP
2. NaI
3. $CH_3CO_2Bu^t$ LDA
HO
OH
1. DIBAL
2. $BuPPh_3^+Br^-$ NaHMDS
3. H_3O^+
HO
OH
$Pd(OAc)_2$
H
steps

Scheme 8.1

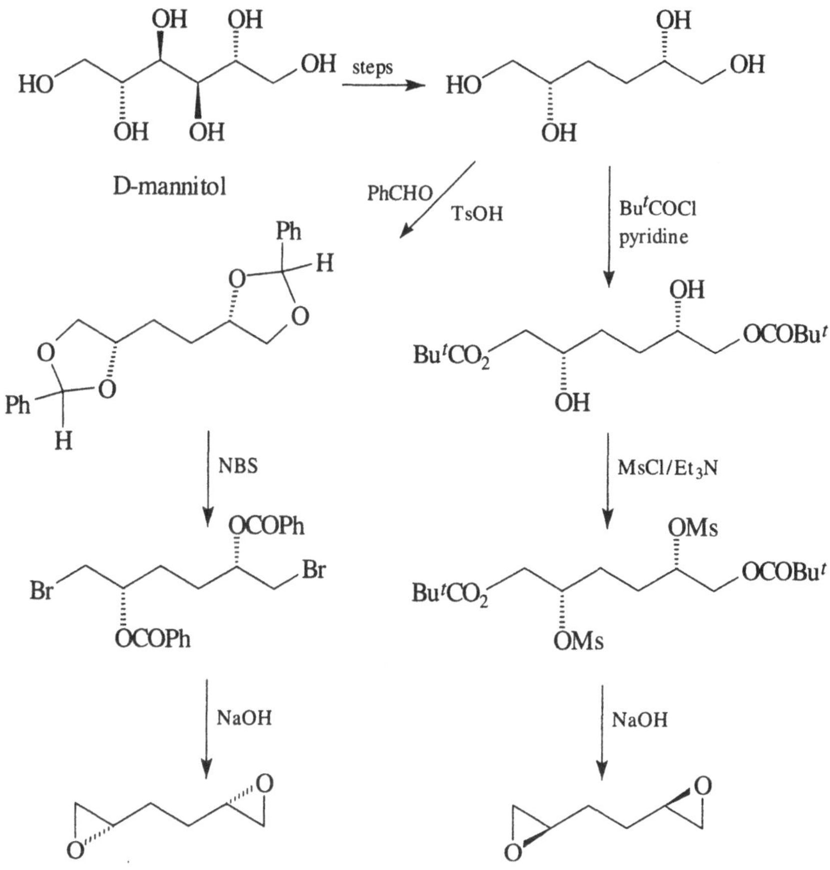

Scheme 8.2

8.2 SYNTHESIS OF MOLECULES WITHOUT C$_2$ SYMMETRY

One simple way in which molecules lacking C$_2$ symmetry can be synthesized is by symmetrical cleavage of a compound with C$_2$ symmetry (Scheme 8.4).

More usually desymmetrization is achieved by making a change to one functional group. It is not always easy to carry out a reaction on only one functional group in a bifunctional compound, but since the starting material is chiral the process proceeds under substrate control and is stereoselective. For example, the *bis*-epoxides prepared in Scheme 8.2 can be used as starting materials for the synthesis of a series of enantiomeric indolizidine alkaloids (Scheme 8.5).

(+)-Diisopropyl tartrate has been used as a starting material for the synthesis of (−)-hikizimycin (Scheme 8.6). In this, as in other examples, it is convenient

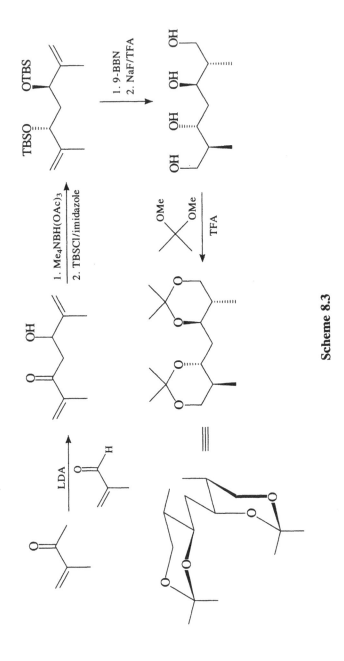

Scheme 8.3

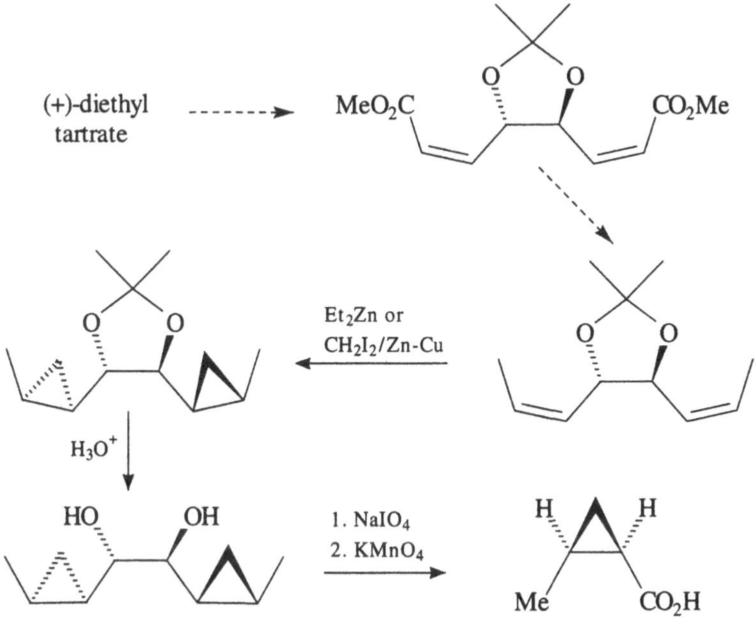

Scheme 8.4

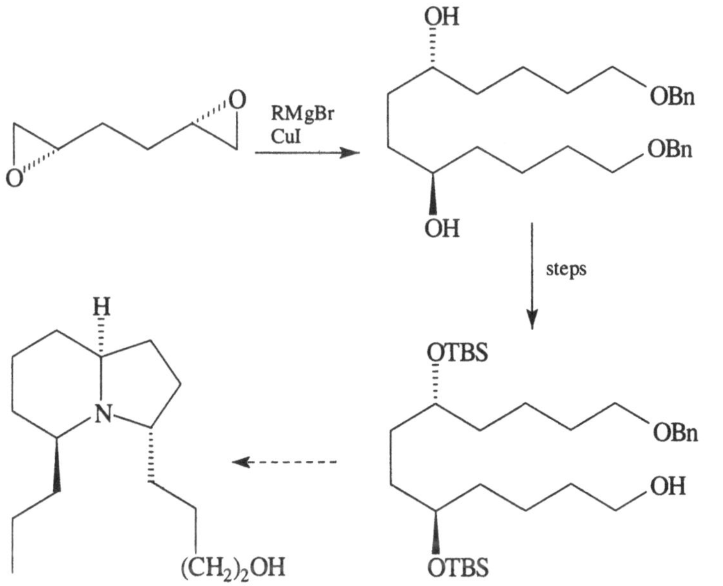

Scheme 8.5

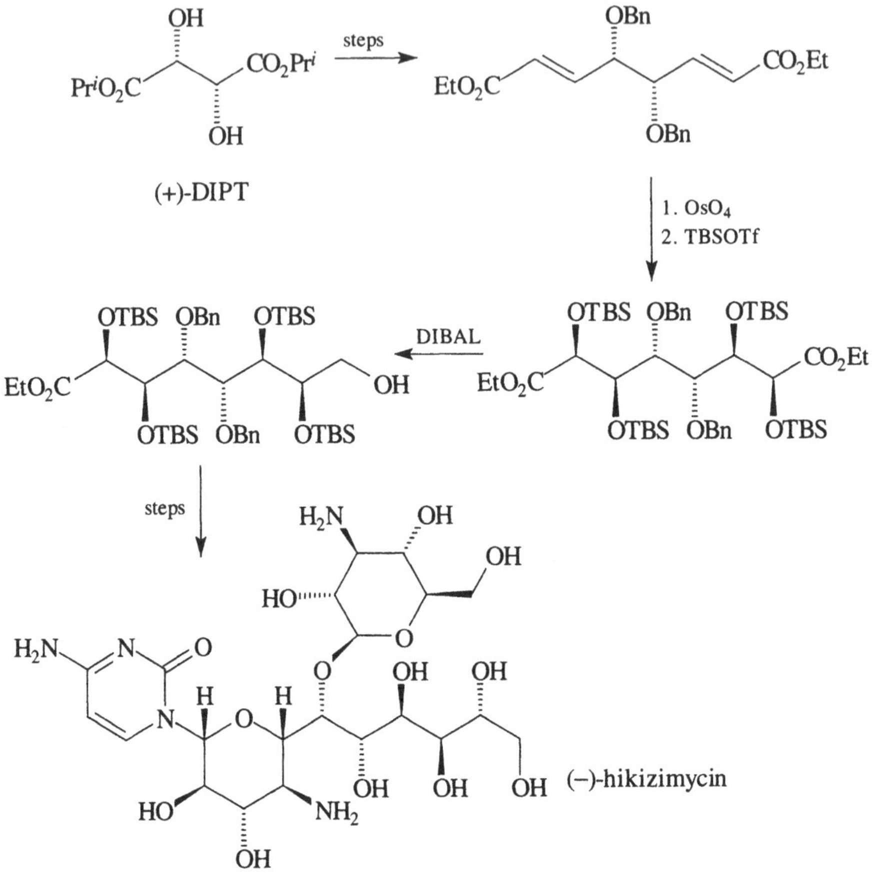

Scheme 8.6

to retain the C_2 symmetry to extend the chain stereoselectively in both directions. Terminus differentiation is then achieved by reduction of one of the two ester groups using diisobutylaluminium hydride.

Nakamura *et al.* have utilized a C_2 symmetric epoxide and spiroketal to synthesize tetranormethyl calcimycin (Scheme 8.7). Terminus differentiation is in this case achieved by protection of one of the primary OH groups in the spiro-ketal.

Barrett *et al.* have used enantioselective cyclopropanation to construct the lipophilic side-chain of an anti-fungal nucleoside isolated from the fermentation broth of *Streptoverticillium fervens* (Scheme 8.8). Monoprotection of the diol destroys the C_2 symmetry, leading eventually to the unsymmetrically substituted product.

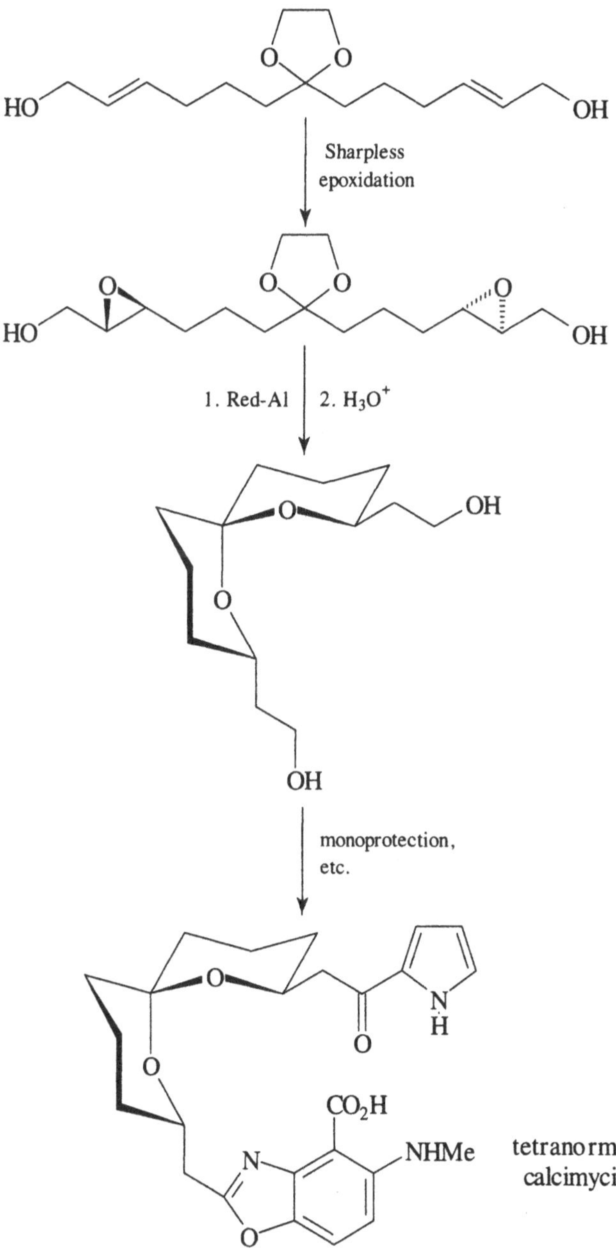

Scheme 8.7

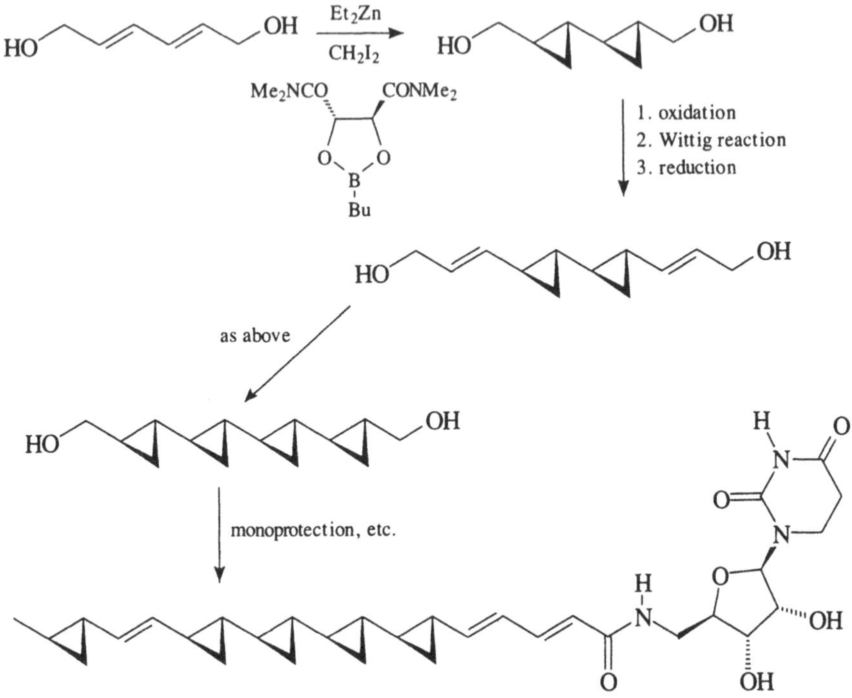

Scheme 8.8

The early stages of a synthesis of mycoticin A are shown in Scheme 8.9. While it is convenient to retain the C$_2$ symmetry in the early stages of the synthesis, a simple monoprotection is all that is needed to produce an unsymmetrical product.

8.3 COMPARISON OF σ SYMMETRIC (*MESO*) AND C$_2$ SYMMETRIC COMPOUNDS

It is instructive to compare the reactions of *meso* and racemic C$_2$ symmetric compounds with chiral reagents. For example, the reaction of a cyclic anhydride with a chiral alcohol leads to loss of symmetry in one case and kinetic resolution in the other (eqns 8.1 and 8.2).

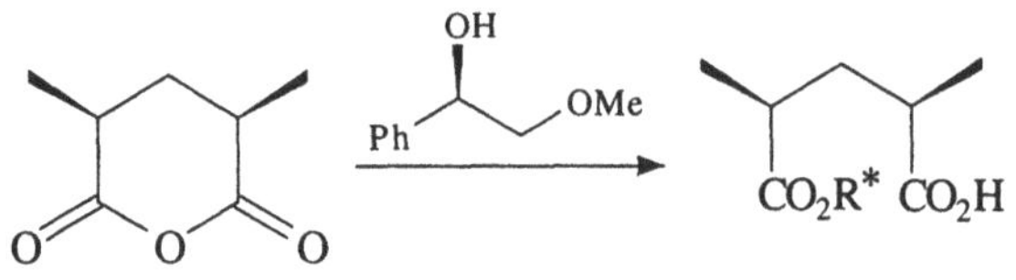

(8.1)

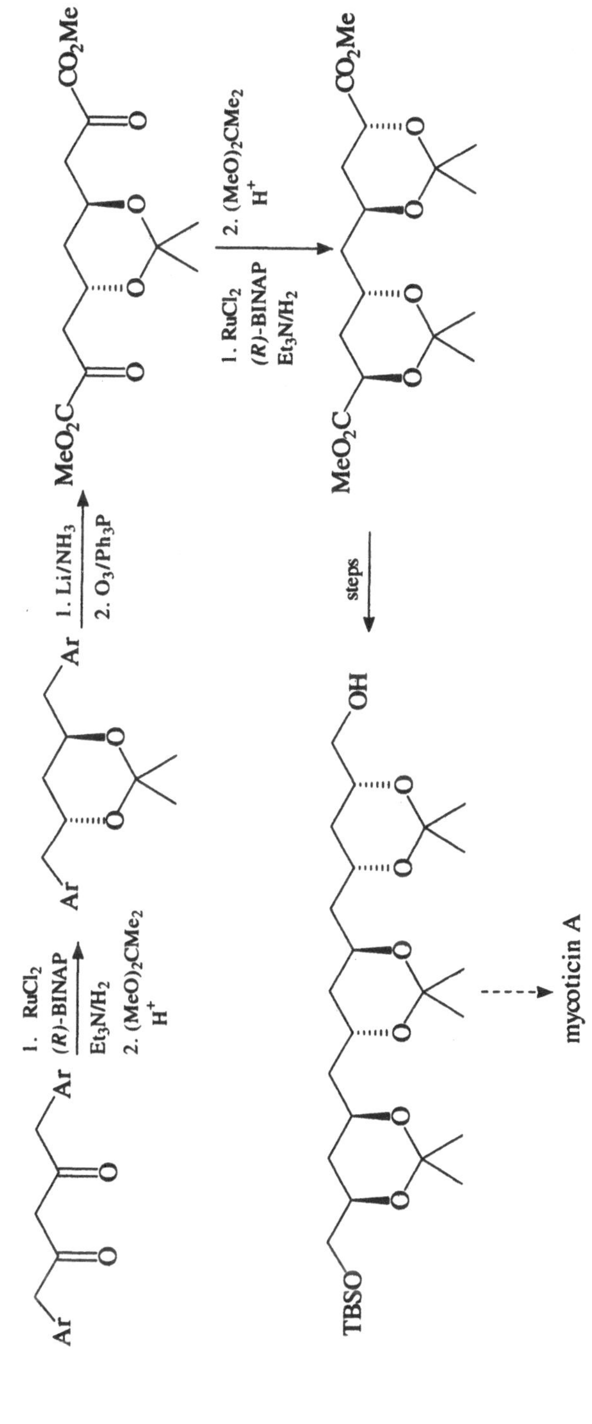
Ar
Ar
1. RuCl2
(R)-BINAP
Et3N/H2
2. (MeO)2CMe2
H+
Ar
Ar
1. Li/NH3
2. O3/Ph3P
MeO2C
CO2Me
1. RuCl2
(R)-BINAP
Et3N/H2
2. (MeO)2CMe2
H+
MeO2C
CO2Me
steps
TBSO
OH
mycoticin A

Scheme 8.9

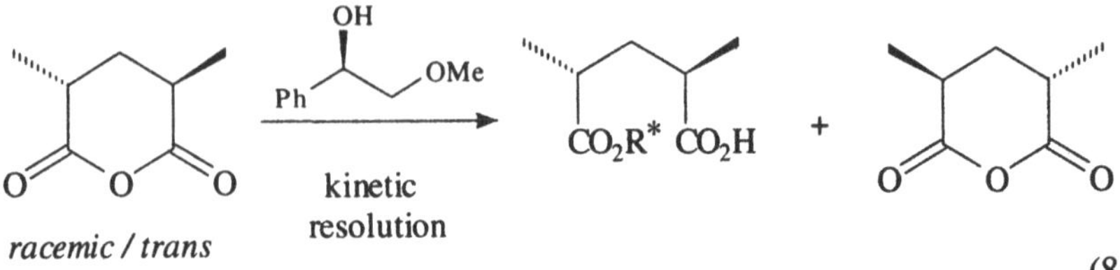

(8.2)

Similarly, reaction with a chiral amine followed by further steps can afford direct access to all three enantiomers of the corresponding lactone (Scheme 8.10).

A further example involves the use of an asymmetric Horner–Emmons reaction using a chiral non-racemic phosphonamide (eqns 8.3 and 8.4). Of course a maximum realistic yield of either of the kinetic resolution products is likely to be around 50%.

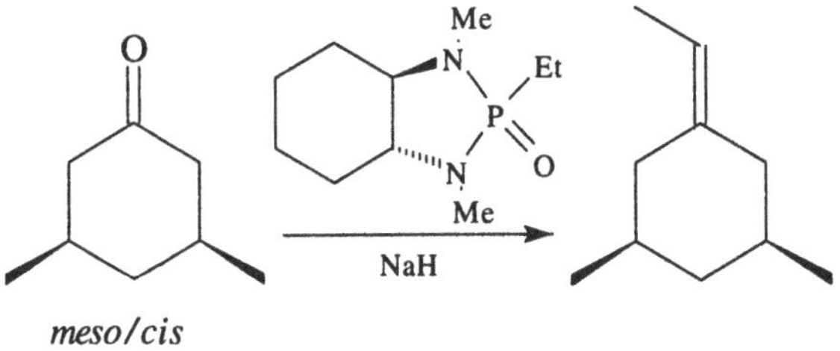

(8.3)

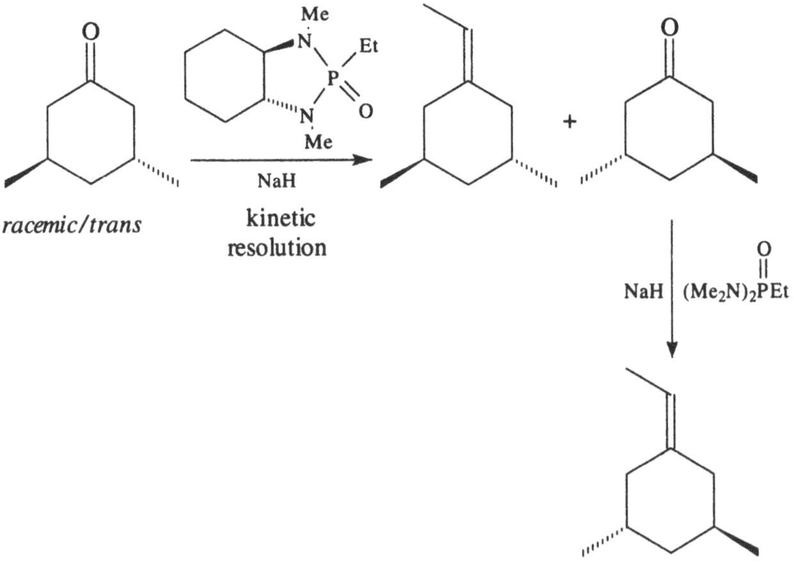

(8.4)

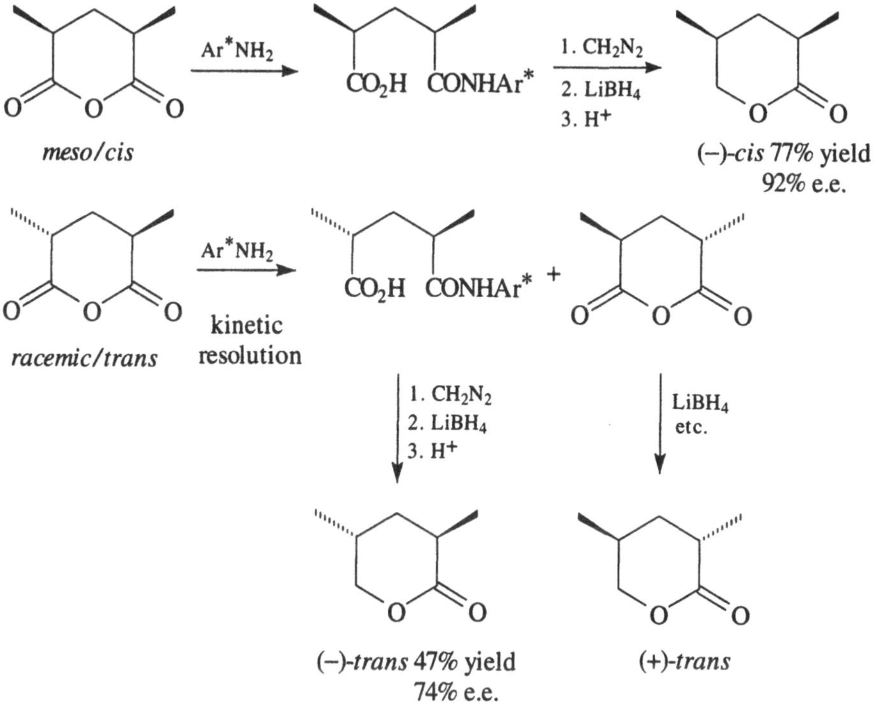

Scheme 8.10

Enantioselective deprotonation of ketones and epoxides provide further illustrations (eqns 8.5–8.8).

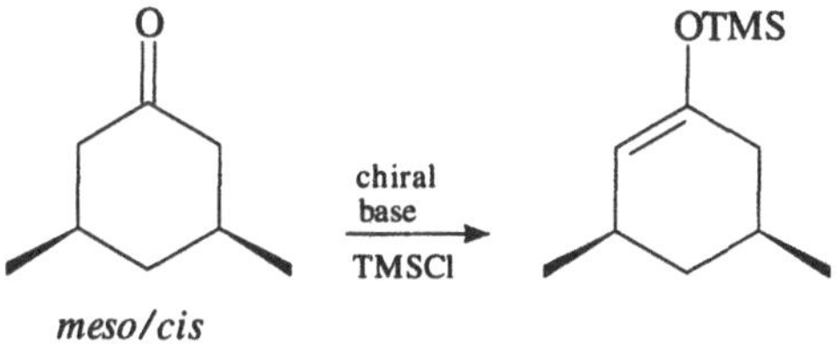

$$(8.5)$$

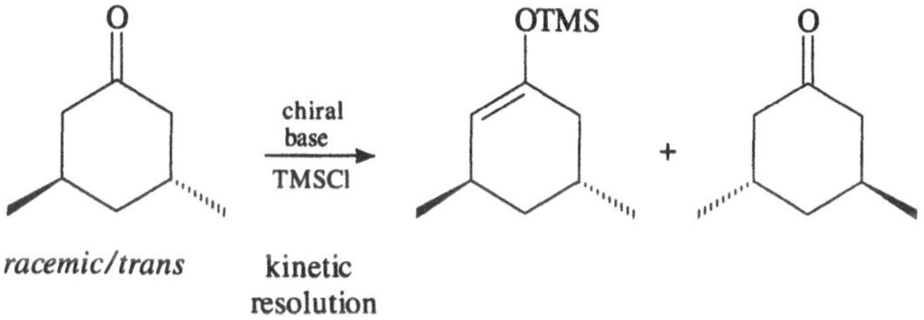

$$(8.6)$$

(8.7)

meso/cis

(8.8)

racemic/trans   kinetic
resolution

The Sharpless epoxidation of allylic/homoallylic diols also make a striking comparison (eqns 8.9 and 8.10).

(8.9)

meso

(8.10)

racemic

8.4 REACTIONS OF MOLECULES WITH PSEUDO C_2 SYMMETRY

The stereochemical relationship between *meso*, C_2 symmetric, and pseudo C_2 symmetric molecules is illustrated in Figure 8.1.

Desymmetrization of pseudo C_2 symmetric compounds involves diastereotopic group selectivity. Hoye *et al.* have studied the reduction of a C_2 symmetric keto-acid to give a pseudo C_2 symmetric hydroxy-acid which cyclizes on acidification. Under kinetic conditions a high degree of stereoselectivity is achieved (Scheme 8.11). All of the substituents on the ring are equatorial in the major product.

A similar tactic has been used in the steps carried out by Schreiber *et al.* as part of a synthesis of FK506 (Scheme 8.12).

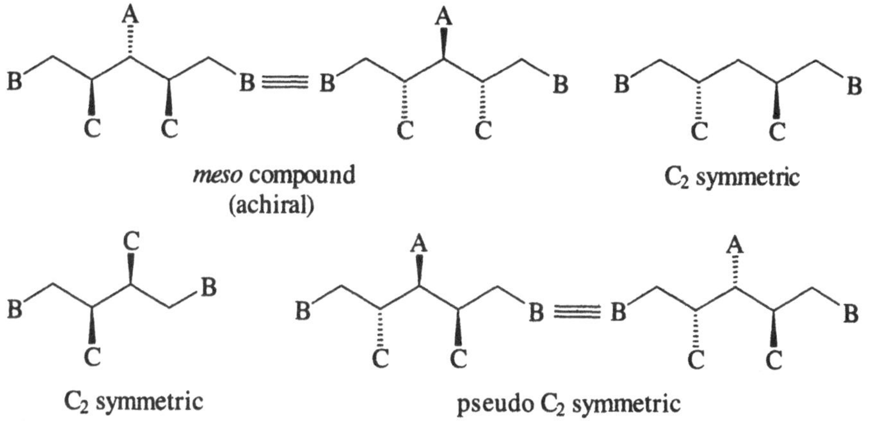

Figure 8.1

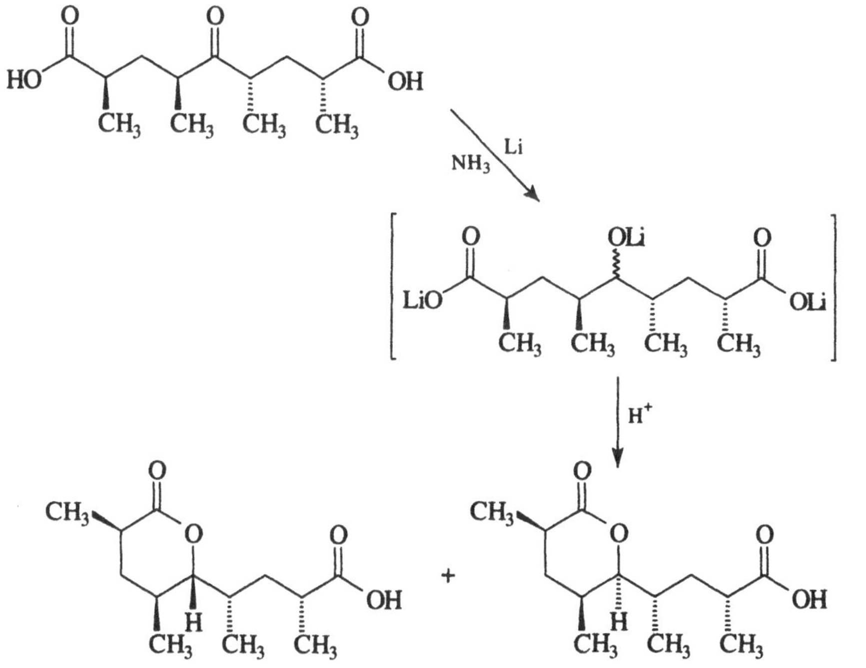

kinetic ratio	350	:	1	(pH 3)
thermodynamic ratio	4·2	:	1	(pH 1)

Scheme 8.11

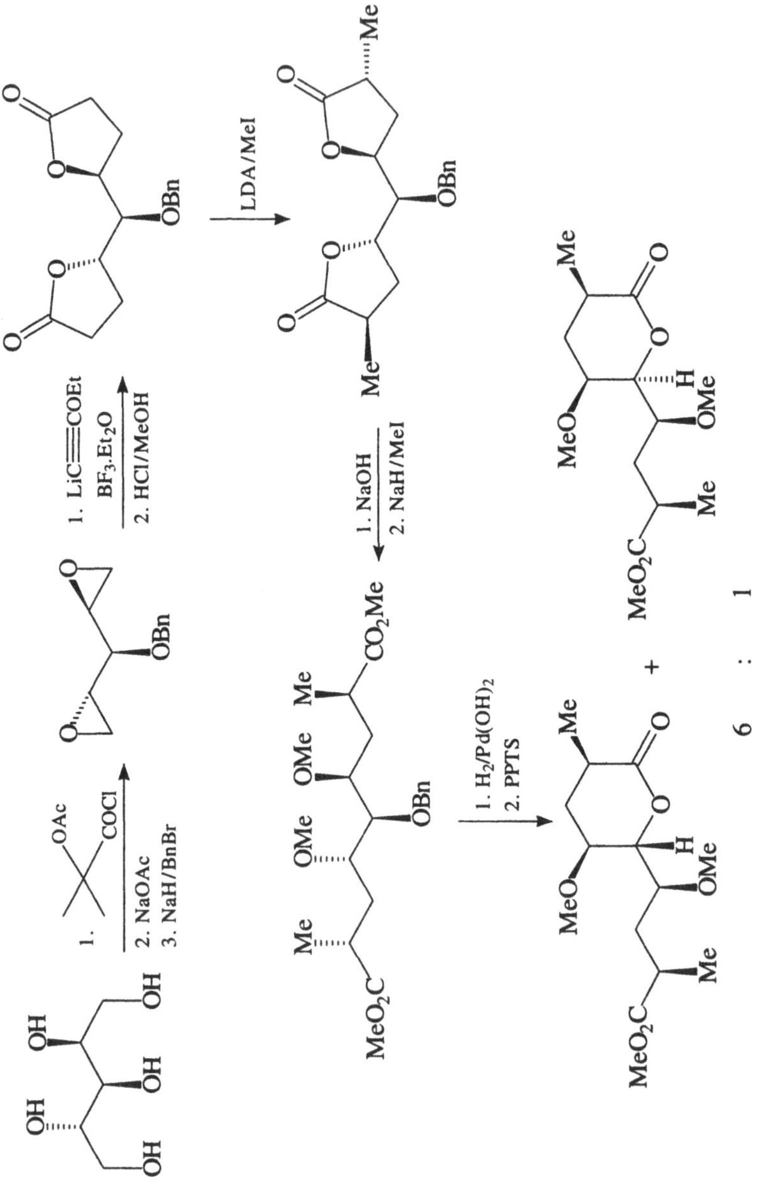
1. OAc COCl
2. NaOAc
3. NaH/BnBr
1. LiC≡COEt
BF3.Et2O
2. HCl/MeOH
LDA/MeI
1. NaOH
2. NaH/MeI
1. H2/Pd(OH)2
2. PPTS
6 : 1

Scheme 8.12

8.5 IODOLACTONIZATION

Iodolactonization is a highly diastereoselective reaction. The *meso* isomers of unsaturated acids contain enantiotopic groups and diastereotopic faces and display facial selectivity (Scheme 8.13).

The mechanism of the cyclization step is shown in Figure 8.2.

The corresponding pseudo C_2 symmetric isomers contain diastereotopic groups and diastereotopic faces and as a result display both group and facial selectivity (Scheme 8.14).

8.6 CASCADE REACTIONS OF DI- AND TRI-EPOXIDES

Hoye *et al.* have studied a series of cascade reactions of epoxides to form tetrahydrofurans. The overall reaction sequence is shown below (eqn. 8.11).

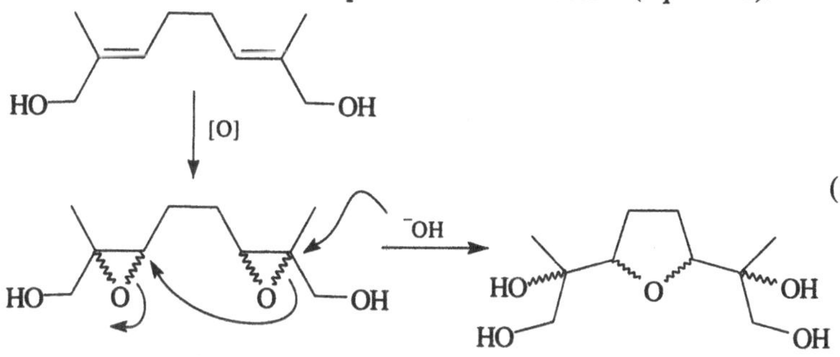

$$(8.11)$$

Meta-chloroperbenzoic acid oxidation gave a mixture of all three diastereomeric epoxides, which with alkali were converted into a mixture of the three stereoisomeric tetrahydrofurans. However, labelling experiments using $H_2{}^{18}O$ confirmed that the *meso* isomer gave rise to the racemic tetrahydrofuran, while the racemic epoxide gave rise to the *meso* tetrahydrofuran (Scheme 8.15). In contrast, Sharpless epoxidation gave either the (+)- or (−)-epoxide which with alkali gave the *meso* tetrahydrofuran. The latter route therefore constitutes an asymmetric synthesis of an achiral (meso) compound!

This strategy can also be extended to the preparation of 2,5-linked *bis*-tetrahydrofurans (Scheme 8.16). Unfortunately, since the cascade reaction can take place either from top to bottom or from bottom to top, and the terminal

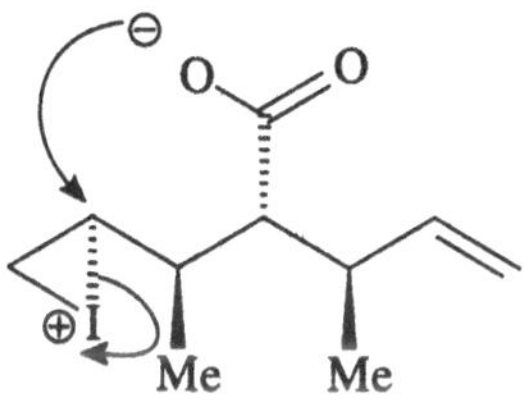

Figure 8.2

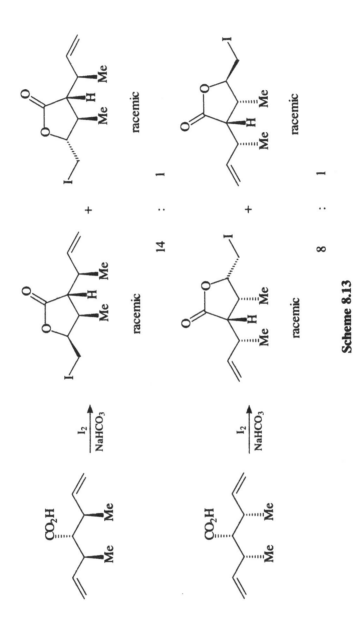

Scheme 8.13

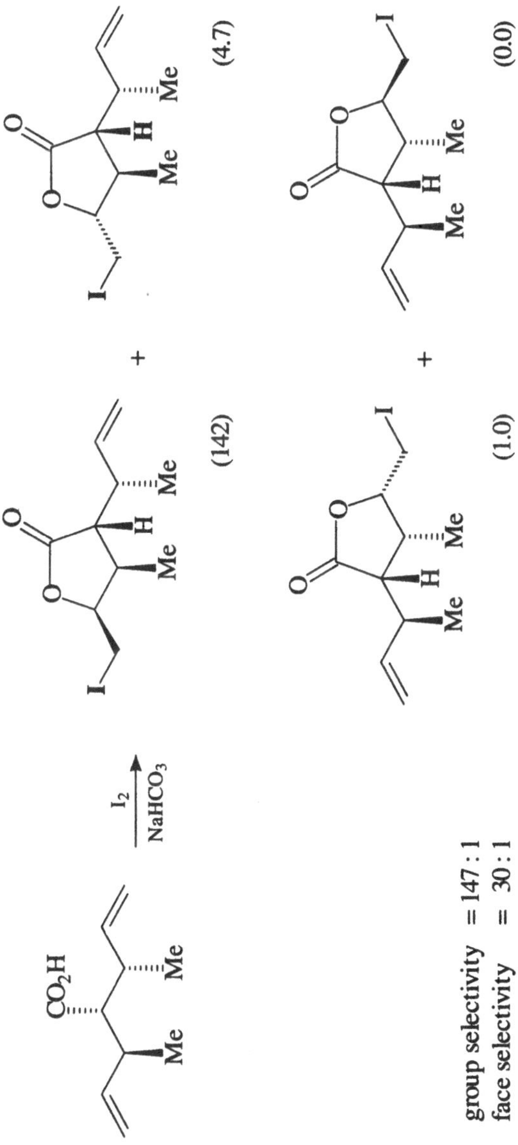

Scheme 8.14

group selectivity = 147 : 1
face selectivity = 30 : 1

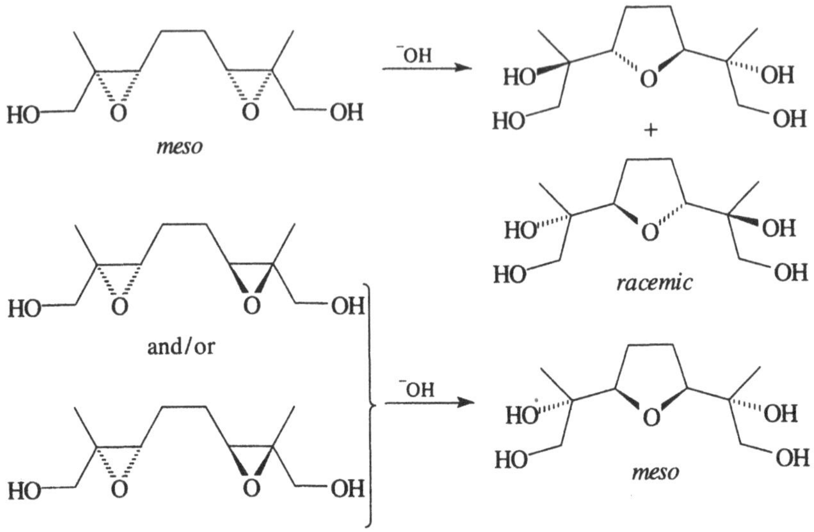

Scheme 8.15

epoxide group is opened by hydroxide, a racemic product is obtained, despite the fact that the enantiomeric excess of the *bis*-epoxide is 99.5%.

However, by adopting an "inside-out" strategy the individual isomeric 2,5-linked *bis*-tetrahydrofurans can be stereoselectively prepared (Scheme 8.17, cf. Scheme 7.10).

This strategy has been used to synthesize (+)-uvaricin (eqn. 8.12).

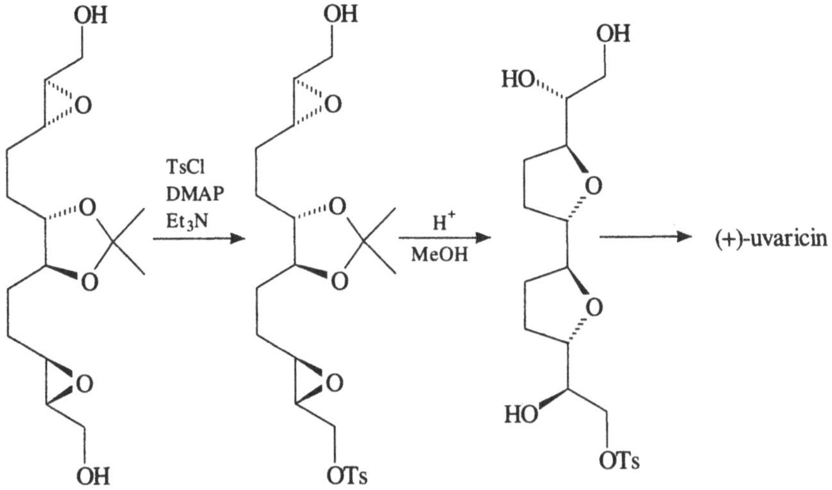

(8.12)

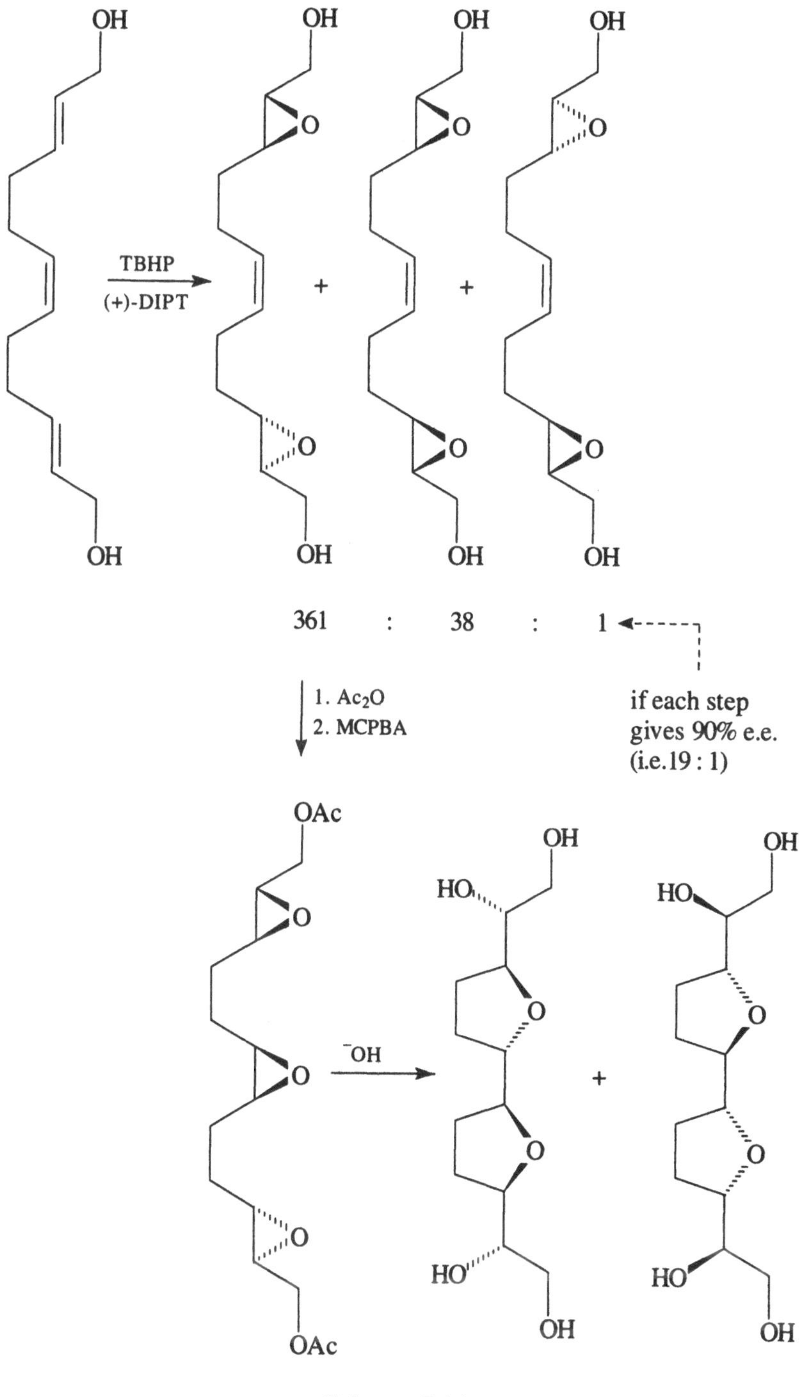
OH
OH
TBHP
(+)-DIPT
+
+
O
361 : 38 : 1
if each step
gives 90% e.e.
(i.e.19 : 1)
1. Ac_2O
2. MCPBA
OAc
^{-}OH
HO

Scheme 8.16

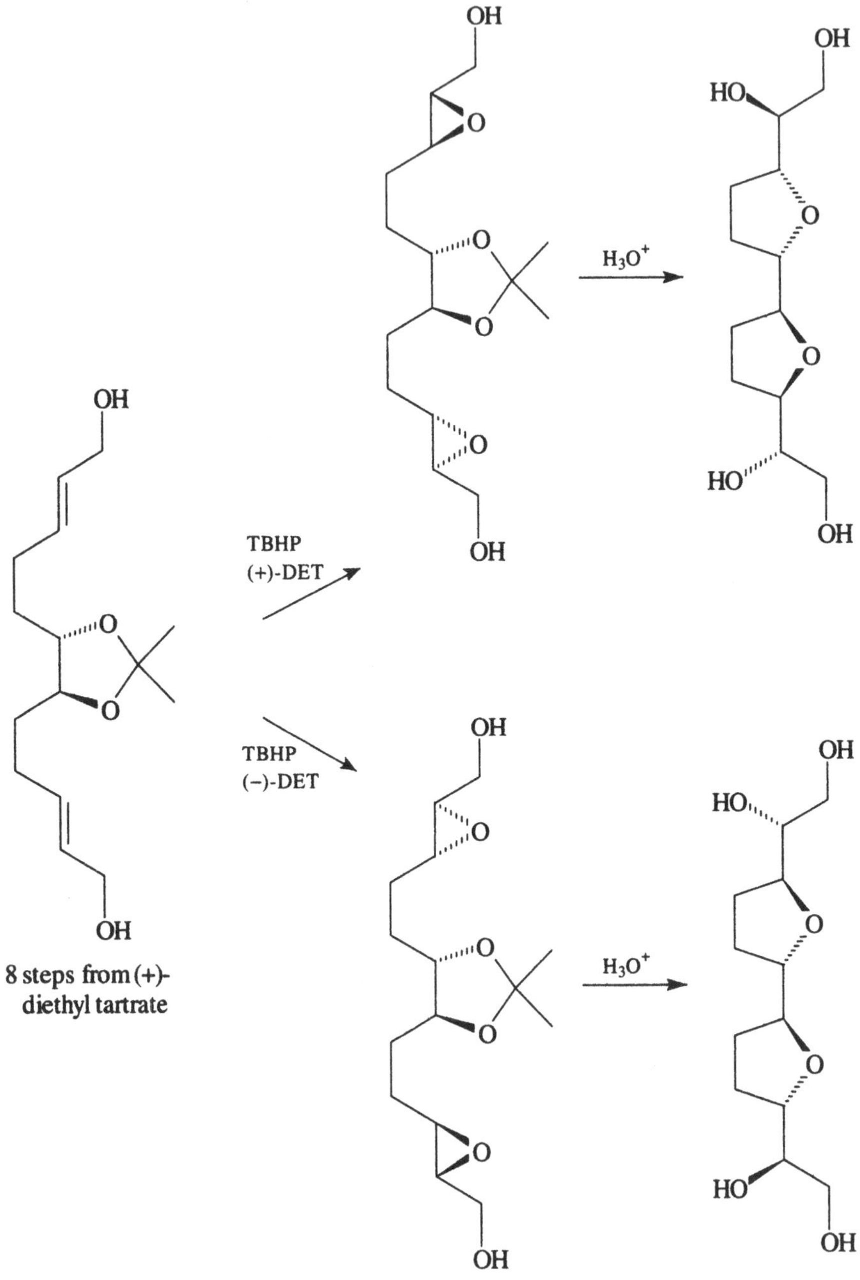
OH
OH
8 steps from (+)-
diethyl tartrate
TBHP
(+)-DET
TBHP
(−)-DET
H_3O^+
H_3O^+
OH
HO
O

Scheme 8.17

PROBLEM

1. Asymmetric epoxidation and dihydroxylation combine to afford an elegant entry to 2,5-linked tetrahydrofurans. Predict the product of the acid-catalysed cyclization.

FURTHER READING

C. S. Poss and S. L. Schreiber, *Acc. Chem. Res.*, 1994, *27*, 9.
S. R. Magnusson, *Tetrahedron*, 1995, *51*, 2167.
J. K. Whitesell, *Chem. Rev.*, 1989, *89*, 1581.

Synthesis of a functionalised ionophore

M. T. Burger and W. C. Still, *J. Org. Chem.*, 1996, *61*, 775.

Synthesis of hikizimycin

N. Ikemoto and S. L. Schreiber, *J. Am. Chem. Soc.*, 1990, *112*, 9657 and 1992, *114*, 2524.

Synthesis of tetranormethyl calcimycin

Y. Nakahara, A. Fujita and T. Ogawa, *Agric. Biol. Chem.*, 1987, *51*, 1009.

Synthesis of polycyclopropanes

A. G. M. Barrett and K. Kasdorf, *J. Chem. Soc., Chem. Commun.*, 1996, 325.
A. G. M. Barrett, W. W. Doubleday, K. Kasdorf and G. J. Tustin, *J. Org. Chem.*, 1996, *61*, 3280.
A. G. M. Barrett, D, Hamprecht, A. J. P. White and D. J. Williams, *J. Am. Chem. Soc.*, 1996, *118*, 7863.

Synthesis of FK506

S. L. Schreiber, T. Sammakia and D. E. Uehling, *J. Org. Chem.*, 1989, *54*, 15.
M. Nakatsuka, J. A. Ragan, T. Sammakia, D. B. Smith, D. E. Uehling and S. L. Schreiber, *J. Am. Chem. Soc.*, 1990, *112*, 5583.

Reactions of *bis*-epoxides

N. Machinaga and C. Kibayashi, *J. Org. Chem.*, 1992, *57*, 5178.
R. E. Babine, N. Zhang, A. R. Jurgens, S. R. Schow, P. R. Desai, J. C. James and M. F. Semmelhack, *Bioorg. & Med. Chem. Lett.*, 1992, *2*, 541.

Synthesis of *bis*-tetrahydrofurans

T. R. Hoye, D. R. Peck and P. K. Trumper, *J. Am. Chem. Soc.*, 1981, *103*, 5618.
T. R. Hoye, D. R. Peck and T. A. Swanson, *J. Am. Chem. Soc.*, 1984, *106*, 2738.
T. R. Hoye and J. C. Suhadolnik, *J. Am. Chem. Soc.*, 1985, *107*, 5312; *Tetrahedron*, 1986, *42*, 2855.
T. R. Hoye and S. A. Jenkins, *J. Am. Chem. Soc.*, 1987, *109*, 6196.
J. A. Marshall and K. W. Hinkle, *J. Org. Chem.*, 1996, *61*, 4247.

ANSWER

1.

(cf. T. R. Hoye and Z. Ye, *J. Am. Chem. Soc.*, 1996, *118*, 1801.)

9

Dynamic Resolution Procedures

Kinetic resolution has long been recognized as an effective tool for the preparation of enantiomerically enriched compounds. Like conventional resolution processes, however, the maximum yield of one stereoisomer of the starting material or product which can be obtained is 50%. Therefore any procedure which allows epimerization of the substrate prior to the reaction has the advantage that it can in principle lead to quantitative conversion of a racemic starting material into a single stereoisomer of the product.

9.1 DYNAMIC KINETIC RESOLUTION (DKR)

The simplest process of this type involving the equilibration of two enantiomers is shown below (Scheme 9.1). For the process to be effective the rate of racemization (k_{rac}) must be at least equal to (or faster than) the rate of reaction of the faster reacting enantiomer (k_S).

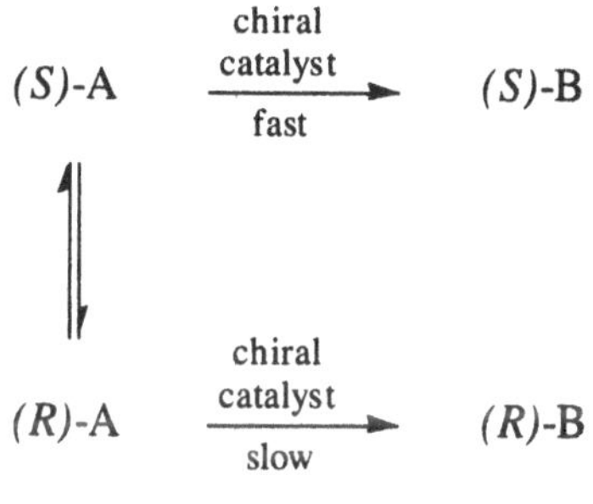

$$(S)\text{-A} \quad \xrightarrow[\text{fast}]{\substack{\text{chiral} \\ \text{catalyst}}} \quad (S)\text{-B}$$

$$\updownarrow$$

$$(R)\text{-A} \quad \xrightarrow[\text{slow}]{\substack{\text{chiral} \\ \text{catalyst}}} \quad (R)\text{-B}$$

Scheme 9.1

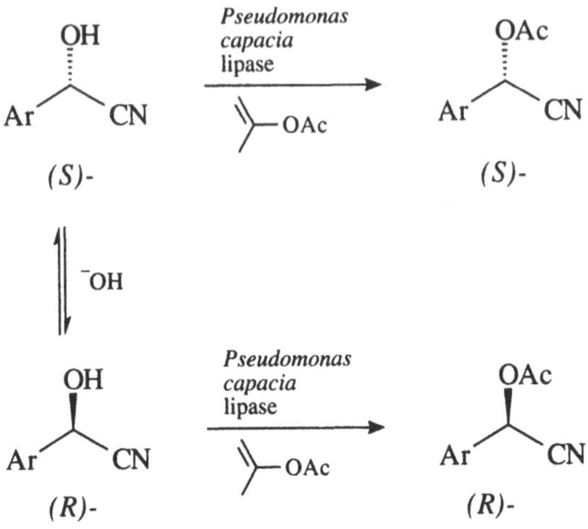

Scheme 9.2

This type of reaction is exemplified by the enzyme-catalysed acetylation of a racemic cyanohydrin (Scheme 9.2).

When the reaction is carried out in the presence of a basic anion exchange resin rapid interconversion of the (R)- and (S)-enantiomers occurs. In practice the parent aldehyde can be converted into one enantiomer of the cyanohydrin acetate in high yield. For example, benzaldehyde gives a 96% yield of the (S)-cyanohydrin acetate with 84% e.e.

A number of other transformations of this type are shown in eqns 9.1–9.7.

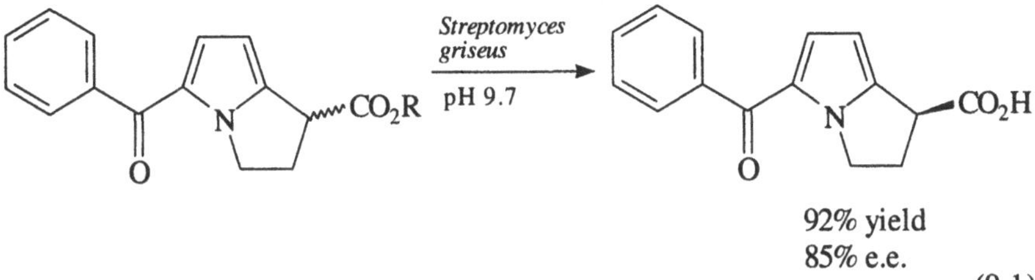

92% yield
85% e.e.

(9.1)

100% yield
76% e.e.

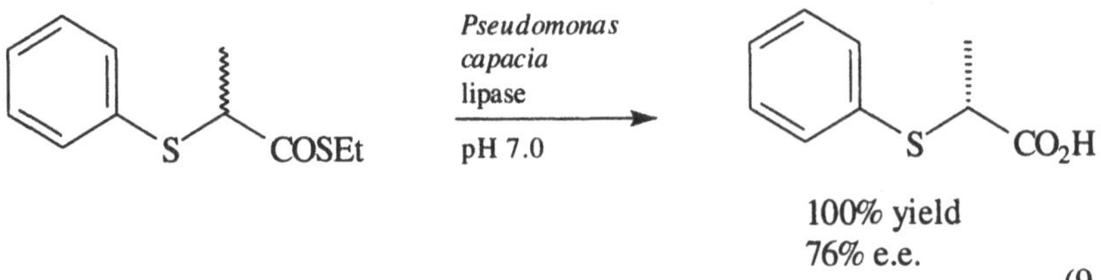

(9.2)

$$
\text{(9.3)} \quad 91\% \text{ yield} \quad \sim 100\% \text{ e.e.}
$$

$$
\begin{array}{ll}
R = \text{butyl} & 87\% \text{ yield}, \quad 87\% \text{ e.e.} \\
R = \text{octyl} & 88\% \text{ yield}, >95\% \text{ e.e.}
\end{array}
\quad \text{(9.4)}
$$

$$
78\% \text{ yield} \quad 90\% \text{ e.e.} \quad \text{(9.5)}
$$

$$
75\% \text{ yield} \quad >95\% \text{ e.e.} \quad \text{(9.6)}
$$

$$
72\% \text{ yield} \quad >95\% \text{ e.e.} \quad \text{(9.7)}
$$

A second type of dynamic kinetic resolution is encountered in the reaction of a mixture of equilibrating diastereomers with an achiral reagent (Scheme 9.3). An example of this type is provided by the reaction of a mixture of epimeric α-bromoamides with a nucleophile (Scheme 9.4).

Heating the α-bromoamide in acetonitrile or DMSO causes epimerization. Furthermore this process takes place more rapidly in the presence of additives such as KBr. Reaction with a soft, unhindered nucleophile such as dibenzylamine

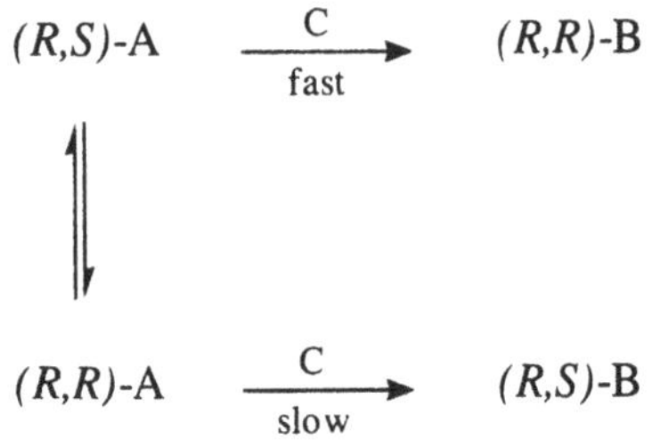

$$(R,S)\text{-A} \xrightarrow[\text{fast}]{\text{C}} (R,R)\text{-B}$$

$$(R,R)\text{-A} \xrightarrow[\text{slow}]{\text{C}} (R,S)\text{-B}$$

Scheme 9.3

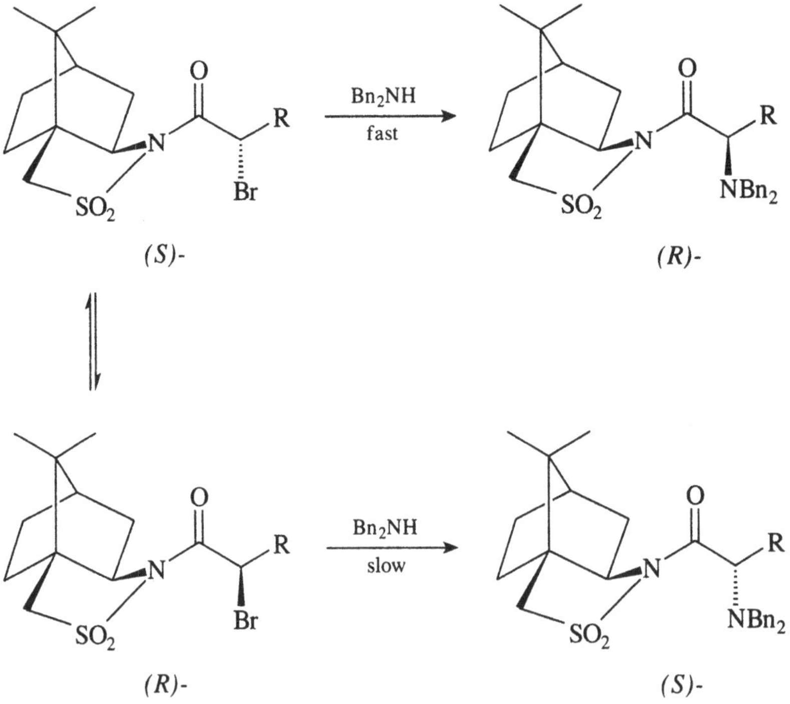

Scheme 9.4

allows equilibration to occur giving a single diastereomer of the product in quantitative yield. In contrast, when a hard, unhindered nucleophile such as sodium azide is used the displacement occurs without epimerization giving the (*R*)-azide from the (*S*)-bromide and *vice versa*. Similar results are obtained with oxygen and sulfur nucleophiles.

The reactions of diastereomeric α-bromoamides with nucleophiles has also been studied by Nunami *et al.* (Scheme 9.5). In dichloromethane kinetic resolution occurred without epimerization. However in DMF, DMSO or HMPA one diastereomer of the product was formed in nearly quantitative yield. Nunami *et al.* attempted to rationalize the observed stereoselectivity in terms of a conformational model (Figure 9.1a). However the major product obtained had the opposite configuration to that which would be predicted on purely steric grounds. They therefore proposed that the ester group actually assists in the delivery of the nucleophile (Figure 9.1b).

Caddick *et al.* have looked at a similar reaction (eqn. 9.8), and Durst *et al.* have achieved dynamic kinetic resolution using esters of the chiral alcohol pantolactone (eqn. 9.9).

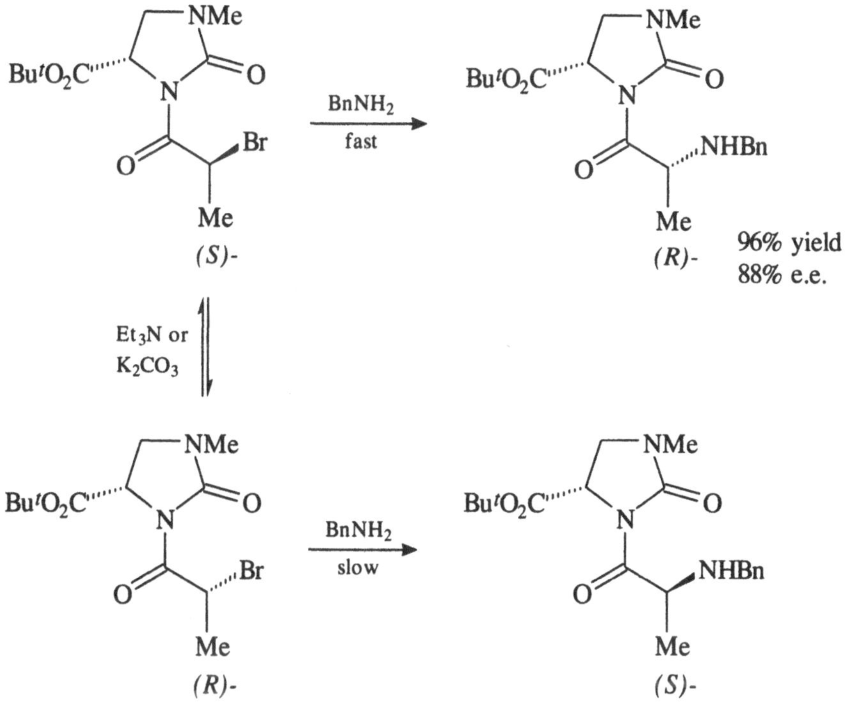

Scheme 9.5

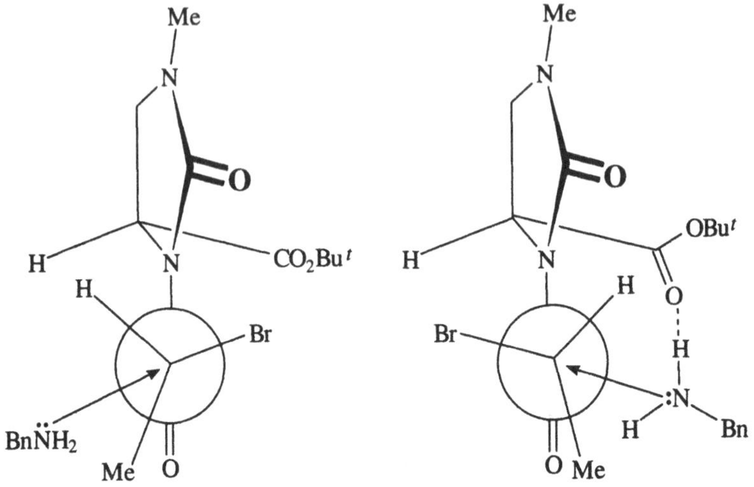

Figure 9.1

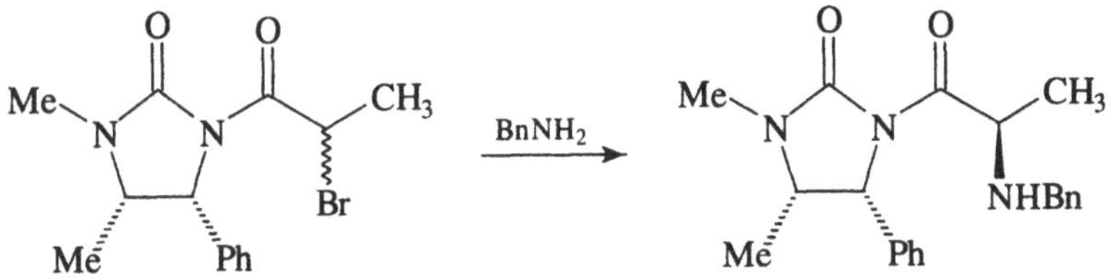

100% yield, 74% e.e.

(9.8)

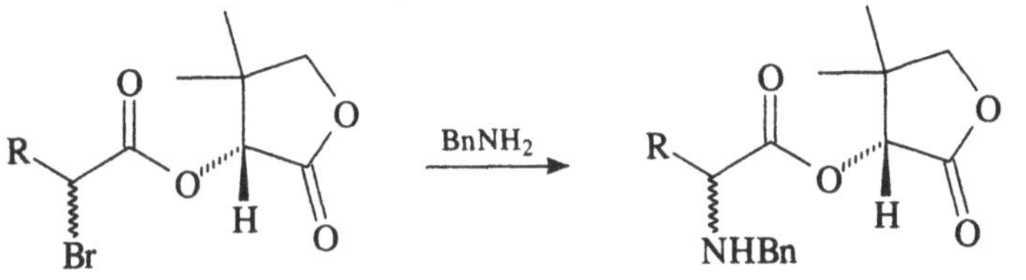

R = Ph   77% yield, 82% e.e.
R = Et   70% yield, 75% e.e.

(9.9)

When the dynamic kinetic resolution occurs along with the creation of a new stereogenic centre the enantioselective synthesis of a compound containing two stereogenic centres is achieved (Scheme 9.6). By selecting appropriate reaction conditions it is then possible to convert a racemic compound into any one of four possible stereoisomers (eqn. 9.10).

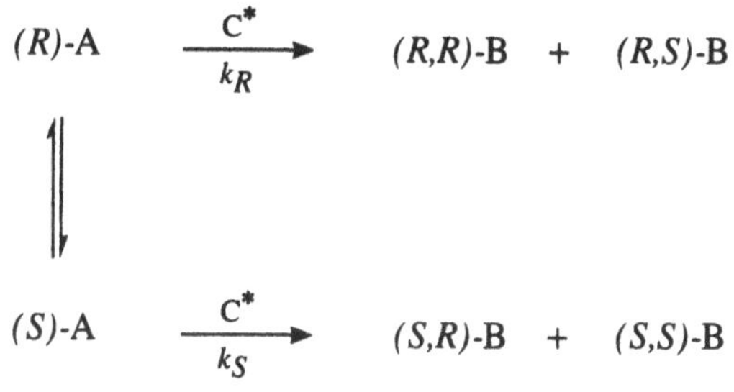

$$(R)\text{-A} \xrightarrow[k_R]{C^*} (R,R)\text{-B} \quad + \quad (R,S)\text{-B}$$

$$(S)\text{-A} \xrightarrow[k_S]{C^*} (S,R)\text{-B} \quad + \quad (S,S)\text{-B}$$

Scheme 9.6

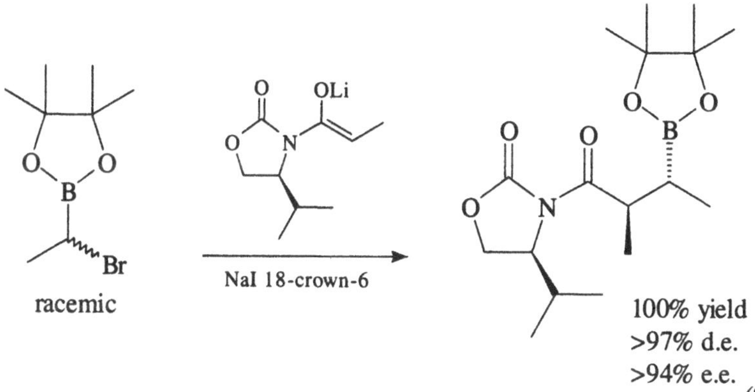

100% yield
>97% d.e.
>94% e.e.
(9.10)

Some of the earliest examples of dynamic kinetic resolution involve stereoselective hydrogenation of β-keto esters using Ni or Ru catalysts (Scheme 9.7).

The reduction of β-keto esters can also be accomplished using Baker's yeast and other microorganisms (eqn. 9.11).

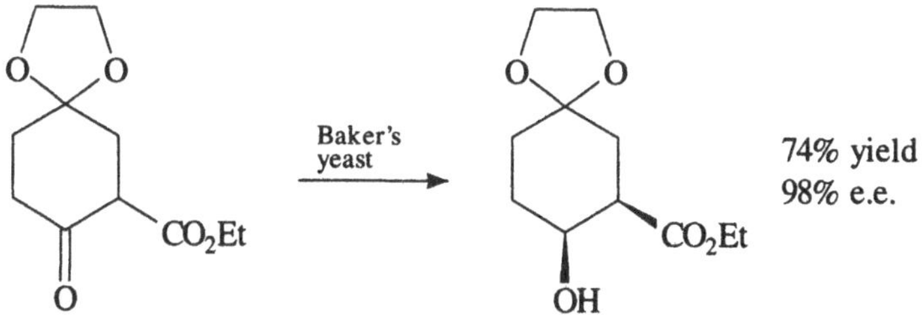

74% yield (9.11)
98% e.e.

By careful selection of the appropriate reaction conditions it is possible to obtain any one of the four possible diastereomers from a given β-keto ester (Scheme 9.8). Enhanced diastereoselectivities can be achieved by using isolated enzymes rather than intact organisms since in Baker's yeast, for example, several competing enzymes may operate.

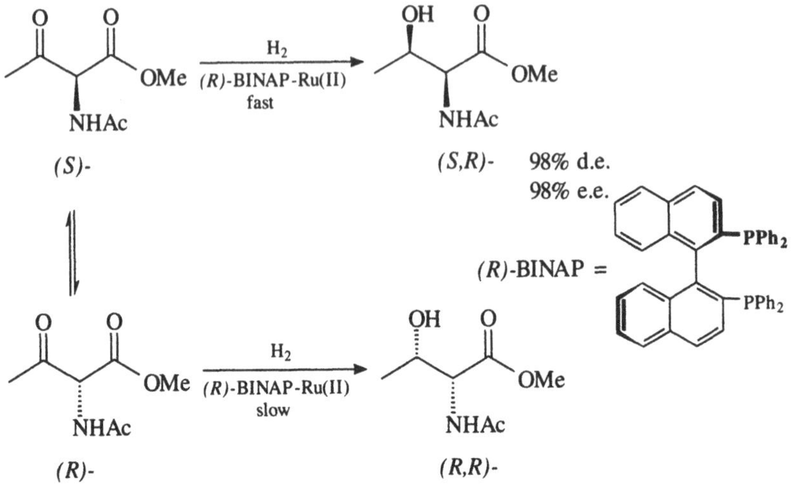

Scheme 9.7

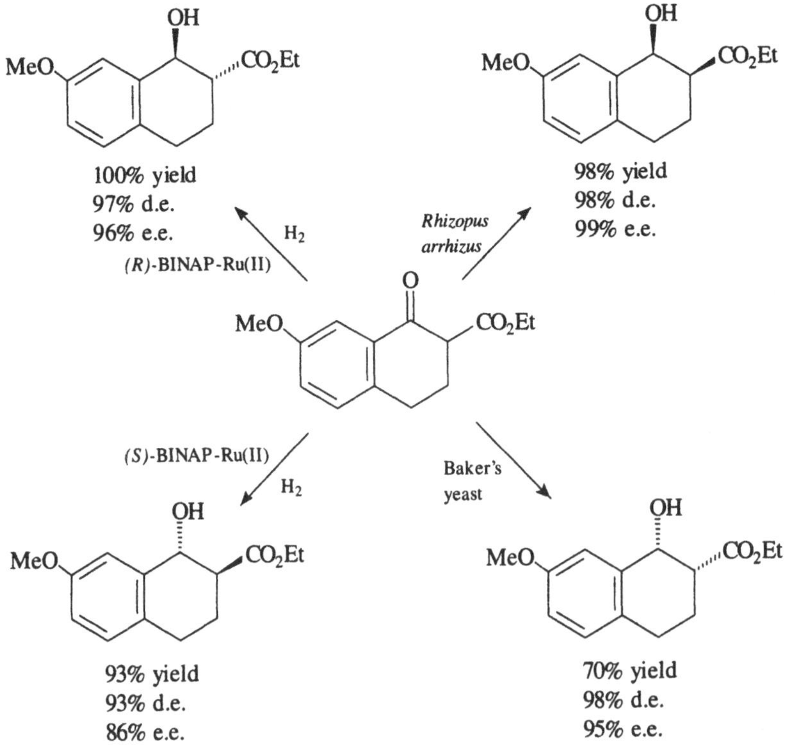

Scheme 9.8

9.2 CRYSTALLIZATION INDUCED DYNAMIC RESOLUTION (CIDR)

Another situation in which a mixture of equilibrating isomers can be converted in high yield into a single isomer arises when one isomer of the starting material or product crystallizes out of solution. For example, racemic narwedine racemizes under basic conditions *via* a retro-Michael reaction. Essentially pure (−)-narwedine can be crystallized in 84% yield when a solution in EtOH/Et$_3$N is seeded with some crystals of the (−)-enantiomer (eqn. 9.12).

$$(9.12)$$

In a second example the benzodiazepine derivative epimerizes in the presence of a catalytic amount of an aromatic aldehyde *via* the more acidic intermediate imine (eqn. 9.13). Addition of (+)-camphor-10-sulfonic acid affords a crystalline salt of the (*S*)-amine with excellent diastereomeric excess. Finally crystallization of a mixture of the epimeric α-bromo-amides in the presence of tetrabutylammonium bromide yields the (*R*)-diastereomer in 91% yield (eqn. 9.14).

91% yield
>98% d.e.

$$(9.13)$$

91% yield
98% d.e.

$$(9.14)$$

PROBLEMS

1. The racemic α-hydroxy-ketone shown below undergoes kinetic resolution on treatment with vinyl acetate in the presence of a lipase. When triethylamine is also present dynamic kinetic resolution occurs to give a high yield of the (−)-acetate. Explain.

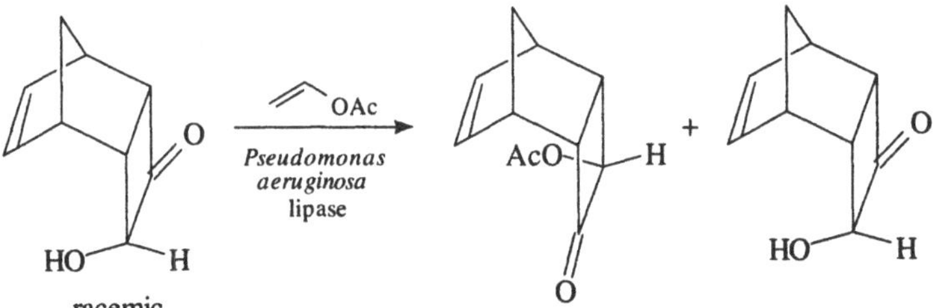

O
HO
H
OAc
Pseudomonas
aeruginosa
lipase
AcO
H
O
+
O
HO
H

racemic

45% yield
>99% e.e.

53% yield
92% e.e.

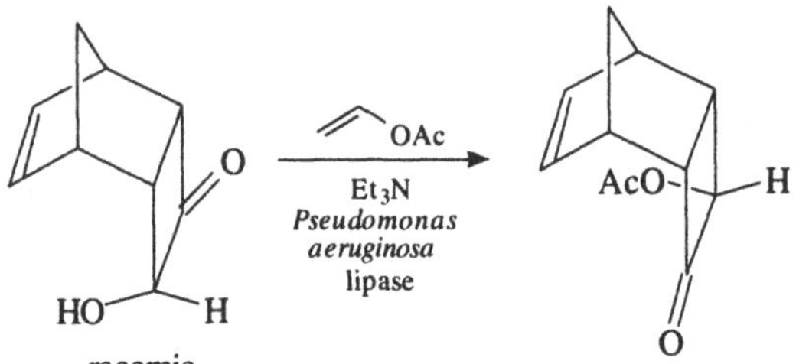

O
HO
H
OAc
Et_3N
Pseudomonas
aeruginosa
lipase
AcO
H
O

racemic

75% yield
97% e.e.

2. α-Dibenzylamino aldehydes (**2**) are valuable intermediates in organic synthesis since they undergo non-chelation controlled addition by organometallic reagents with high diastereoselectivity. They can sometimes be prepared from readily available α-amino acids. However an alternative, more general approach starts from the corresponding, racemic α-bromo acid (**1**) using dynamic kinetic resolution. Outline the steps involved in this route.

Br
R
CO_2H
NBn_2
R
CHO
R'M
NBn_2
R
OH
R'

1 **2** **3**

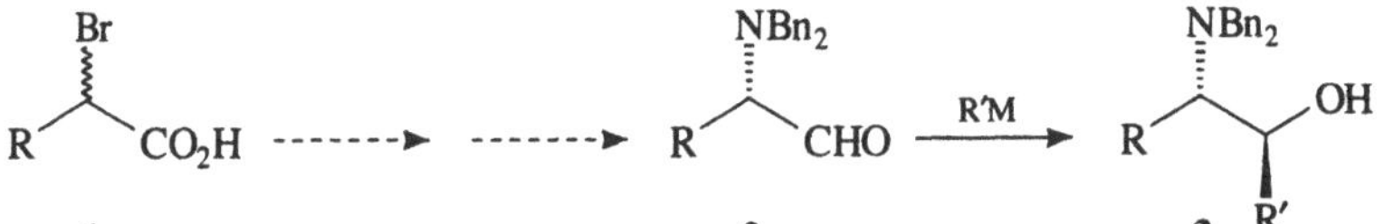

FURTHER READING

R. S. Ward, *Tetrahedron Asymmetry*, 1995, *6*, 1475.
R. Noyori, M. Tokunaga and M. Kitamura, *Bull. Chem. Soc. Jpn.*, 1995, *68*, 36.
R. S. Ward, A. Pelter, D. Goubet and M. C. Pritchard, *Tetrahedron Asymmetry*, 1995, *6*, 469.

K. Nunami, H. Kubota and A. Kubo, *Tetrahedron Lett.*, 1994, *35*, 3107 and 8639.
K. Koh, R. N. Ben and T. Durst, *Tetrahedron Lett.*, 1993, *34*, 4473.
K. Koh and T. Durst, *J. Org. Chem.*, 1994, *59*, 4683.
S. Caddick and K. Jenkins, *Chem. Soc. Rev.*, 1996, *25*, 447.
H. Stecher and K. Faber, *Synthesis*, 1997, 1.

ANSWERS

1. In the presence of triethylamine enolization occurs to give a symmetrical enediol which facilitates equilibration of the enantiomeric α-hydroxy-ketones. Notice that protonation of the enediol leads only to the *endo*-hydroxy compound. The (S)-enantiomer is acetylated most rapidly by the lipase.

(See T. Taniguchi and K. Ogasawara, *J. Chem. Soc., Chem. Commun.*, 1997, 1399.)

2. One possible approach involves reacting the α-bromo acid with an appropriate chiral auxiliary to form an ester or amide derivative which can undergo dynamic kinetic resolution on reaction with dibenzylamine or benzylamine. Removal of the chiral auxiliary leads to a chiral, non-racemic β-amino alcohol or α-amino ester which can be easily converted into the corresponding α-amino aldehyde.

(See J. A. O'Meara, N. Gardee, M. Jung, R. N. Ben and T. Durst, *J. Org. Chem.*, 1998, *63*, 3117.)

Index